HOW TO BUILD ADVANCED SHORT WAVE RECEIVERS

ALSO BY THE SAME AUTHOR

No.222 Solid State Shortwave Receivers For Beginners
No.223 50 Projects Using IC CA3130
No.224 50 CMOS IC Projects

HOW TO BUILD ADVANCED SHORT WAVE RECEIVERS

by

R. A. PENFOLD

BERNARDS (Publishers) LTD
The Grampians
Shepherds Bush Road
London W6 7NF
England

I.S.B.N. 0 900162 67 8

First Published June 1977

Printed and Manufactured in Great Britain by
Hunt Barnard Printing Ltd.

CONTENTS

CHAPTER THREE – DOUBLE CONVERSION RECEIVER

CHAPTER FOUR – ADD-ON CIRCUITS

CHAPTER ONE

ADVANCED S.W. RECEIVERS

Although there is a tendency for short wave listeners and Radio Amateurs to use ready made equipment these days, greater satisfaction and enjoyment can be gained from the hobby by using home constructed equipment. Using ready made S.W. gear does not give any insight into the way in which the apparatus functions, and by building ones own equipment it is virtually inevitable that a reasonable understanding of the techniques involved will be grasped. Such an understanding is very helpful when it comes to actually using a finished receiver, and it enables the operator to obtain optium results from the set.

The reason that commercially produced equipment has increased in popularity is probably largely due to the high standard of finish that is achieved. It must be admitted that it is difficult for the home constructor to equal commercial standards in this respect, but if due care and attention is taken, it is possible to obtain a standard which will satisfy even the most critical.

As far as performance is concerned, there is absolutely no reason why a home constructed receiver should not have a level of performance which is at least equal to that of a commercially built receiver of similar complexity. Furthermore, the home constructed receiver is likely to cost very much less than its ready made equivalent.

Superhet Principle

The receiver circuits which are featured in this book are all of the superheterodyne (usually abbreviated to superhet) type, and circuits of the simpler T.R.F. type will not be considered here. These circuits do not really make good projects for those who have no or only very little experience of S.W. Receiver construction, with the exception of the first project that is described.

Anyone who has tackled a simple T.R.F. S.W. receiver should be able to construct and align this set without too much difficulty, and the primary reason for its inclusion is that it makes a good introduction to superhet designs.

Once some experience has been gained with this set, the constructor should be fully competent to go on and build any of the more advanced designs which are described in subsequent chapters of this book. Apart from providing a good introduction to S.W. superhets, this set also makes a very interesting project in its own right.

Before going on to consider some practical circuits, some background information will be given. It must be stressed that while some of this is of purely an academic nature, some of this is really essential knowledge for anyone undertaking construction of a superhet receiver. This is particularly the case when it comes to the basic principle of the superhet.

Having an understanding of the function of each stage of a superhet makes it easy to put together a practical circuit with confidence. Perhaps more importantly, it takes the guesswork out of aligning and using a finished receiver.

A superhet differs from a straight or T.R.F. receiver in that in the case of the latter, all processing of the signal prior to detection takes place at whatever frequency the signal happens to be at. In the case of a superhet the input signal is converted to an intermediate frequency (I.F.) where it is considerably amplified prior to detection. In a conventional superhet the I.F. is fixed, and no matter what the frequency of the input signal is, it is converted to the intermediate frequency using the heterodyne principle. It is from this that the name superhet is derived.

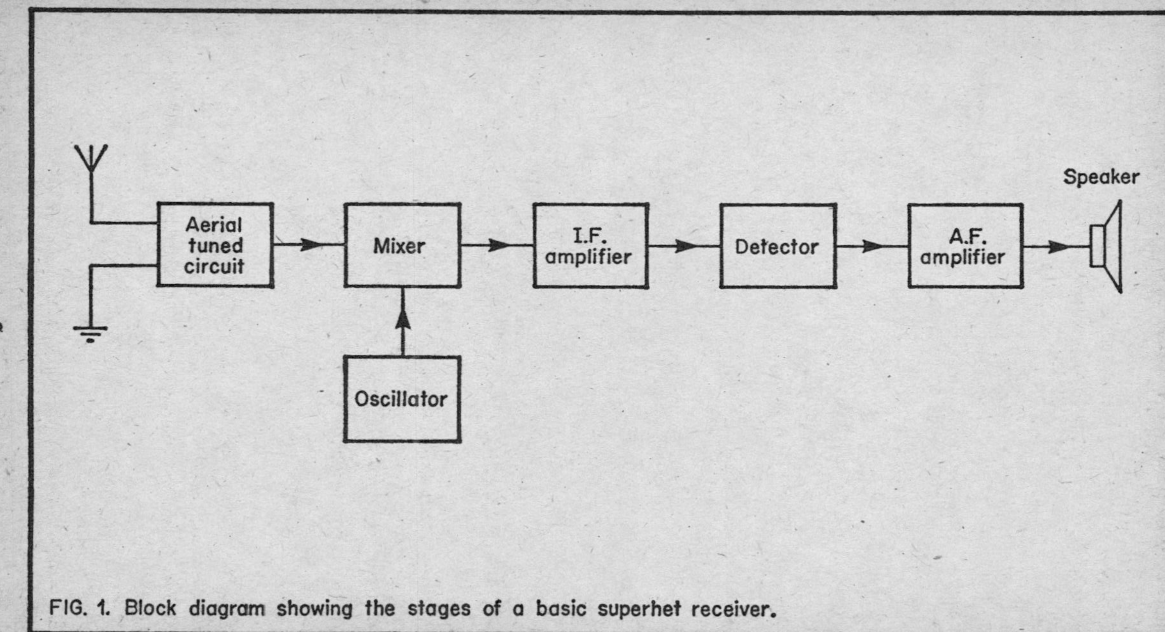

FIG. 1. Block diagram showing the stages of a basic superhet receiver.

A block diagram which illustrates the basic arrangement of a simple superhet is shown in Figure 1. The aerial signal is coupled to a tuned circuit which tunes the desired frequency and rejects all others. The selected signal is then coupled to one input of a mixer, and the other input of the mixer is fed from an oscillator.

There will be four output signals from the mixer, and these are as follows:–

1. The aerial signal.
2. The oscillator signal.
3. The sum of these two frequencies.
4. The difference frequency of the two input signals.

It will be apparent from this that two new frequencies are produced, apart from the original input frequencies. Either of these can be used to provide the I.F. signal, but conventionally it is the difference signal which is used.

A simple mathematical example should help to clarify the function of the mixer and oscillator stages. Suppose that the receiver is required to tune over the range of 5 to 10MHZ, and that an intermediate frequency of 470kHZ is used (the standard intermediate frequency for British transistorised superhets). In order to achieve this it is necessary to arrange the input tuned circuit to tune over a range of 5 – 10MHZ, with the oscillator tuning over a range of 5.47 to 10.47MHZ. In other words the oscillator is always .47MHZ (which is the same as 470kHZ) higher than the input signal frequency. It is not only necessary to have this relationship at band limits, but it must be maintained throughout the tuning range.

Thus if the set is tuned to receive a signal at, say, 7.5MHZ, the oscillator will be at a frequency of 7.97MHZ. This produces the following four output frequencies from the mixer:

1. 7.5MHZ.
2. 7.97MHZ.
3. 15.47MHZ (7.5 + 7.97MHZ).
4. 470kHZ (7.97 - 7.5MHZ).

It is obviously the last of these which is used as the I.F. signal, and the I.F. amplifier will reject the other three signals which are well outside its passband. The I.F. amplifier provides a considerable level of gain, and unlike a T.R.F. set, only a relatively low level of audio gain is required. The detector is a conventional diode type usually.

At first sight the superhet may appear to be a rather overcomplicated setup which offers little advantage, if any, over a T.R.F. design. In fact this is not the case, and even a fairly simple superhet can usually out perform a good T.R.F. design.

The reasons for this are quite straightforward. The two main requirements of a S.W. receiver are for high sensitivity and good selectivity. In order to obtain high sensitivity it is necessary to employ a high degree of R.F. amplification. In order to obtain good selectivity it is necessary to use several tuned circuits.

This is rather difficult to achieve with a T.R.F. design since every tuned circuit will need a tuning capacitor. This, coupled with the fairly high frequencies involved makes high gain rather difficult to achieve, as extensive screening would be required in order to eliminate stray feedback and the instability which would accompany it. Tuned circuits operating at frequencies in the region of tens of MHZ tend to have rather wide bandwidths, and so even by using a number of tuned circuits a very high degree of selectivity would not be obtained.

These difficulties are overcome by using the superhet technique. High gains are not involved at the signal frequency, since the purpose of the early stages of the set is merely to convert the signal to the I.F., and not to provide any significant gain. Obtaining good stability is not, therefore, too much of a problem. It is only necessary to use a twin gang tuning capacitor, with one section being used for the tuning of the aerial tuned circuit, and the other one for the oscillator tuning.

It is when we come to the I.F. amplifier that the main advantages of the superhet become apparent. The comparatively low frequency means that even using only a few tuned circuits results in quite good selectivity. It is a simple matter to achieve high gain and stability, this again being aided by the comparatively low frequency. It is necessary to screen the tuning coils and their tuning capacitors if a very high level of gain is used, but since this amplifier only operates at a single frequency, the tuning capacitance is provided by a fixed capacitor. The coil and capacitor can therefore be a single component which is contained in a metal screening can.

Another advantage of having this single frequency amplifier is that a very high degree of selectivity can be obtained by the use of crystal, mechanical, or ceramic I.F. filters. A common factor to all three of these types of filter is that they can only be used at a single frequency. This makes them completely unsuitable for use with a T.R.F. receiver.

Drawbacks

As one would probably expect, the superhet is not without a few drawbacks. One minor one is that aerial signals at a frequency of 470kHZ can sometimes break through to the I.F. stages. However, this rarely causes any significant trouble, and it can be eliminated by using a wavetrap in the aerial lead. This merely consists of a simple L – C circuit which has a high impedance at and around the I.F., but which has a low impedance at other frequencies. Thus it does not hinder the S.W. aerial signals, but it blocks the interfering signals at the intermediate frequency.

The main problem with superhets is what is known as the image signal or the image response. Superhet receivers actually operate on two frequencies at once, the main response and the image response. In our previous mathematical example the main response was 7.5MHZ, and was produced by an oscillator frequency of 7.97MHZ and an ordinary I.F. of 470kHZ. The

470kHZ I.F. signal was produced by the oscillator and aerial signals difference frequency (7.97 - 7.5 = 0.47MHZ). If an aerial signal at 8.44MHZ were to break through to the input of the mixer, this would also produce a 470kHZ I.F. signal (8.44MHZ - 7.97MHZ = 0.47MHZ). It is this signal which is the image response.

As may already be apparent, the oscillator does not have to be above the aerial signal frequency, but it can be below the input signal frequency. In either case the difference between the two frequencies is equal to the I.F. It is the current convention that the oscillator frequency is the higher of the two however, and it is thus usually the case that the lower response is the main one while the higher response is the image one.

Of course, although the superhet receives on two frequencies simultaneously, this is not to say that it is equally sensitive on each response. The purpose of the aerial tuned circuit is to peak the main response and reject the image one. When using an ordinary I.F. of 470kHZ the responses are 940kHZ apart (twice the 470kHZ I.F.), and if only a single tuned circuit ahead of the mixer is used, the image response rejection will not be very high. This is simply because a single tuned circuit operating at frequencies of around 20 - 30MHZ has a fairly wide bandwidth, and if it is peaked at the main response it will still provide a fairly easy passage for any signal at the image response.

Below about 20MHZ the bandwidth of a single tuned circuit is usually enough to give a reasonably high level of image rejection, but even at these frequencies things could be better.

Therefore, the basic arrangement of Figure 1 is often modified in various ways in order to give improved image rejection, and in order to improve other features of performance such as selectivity. This aspect will be covered in subsequent chapters, but the remainder of this chapter will be concerned with the circuit operation and construction of a simple S.W. superhet receiver based on the simple arrangement of Figure 1.

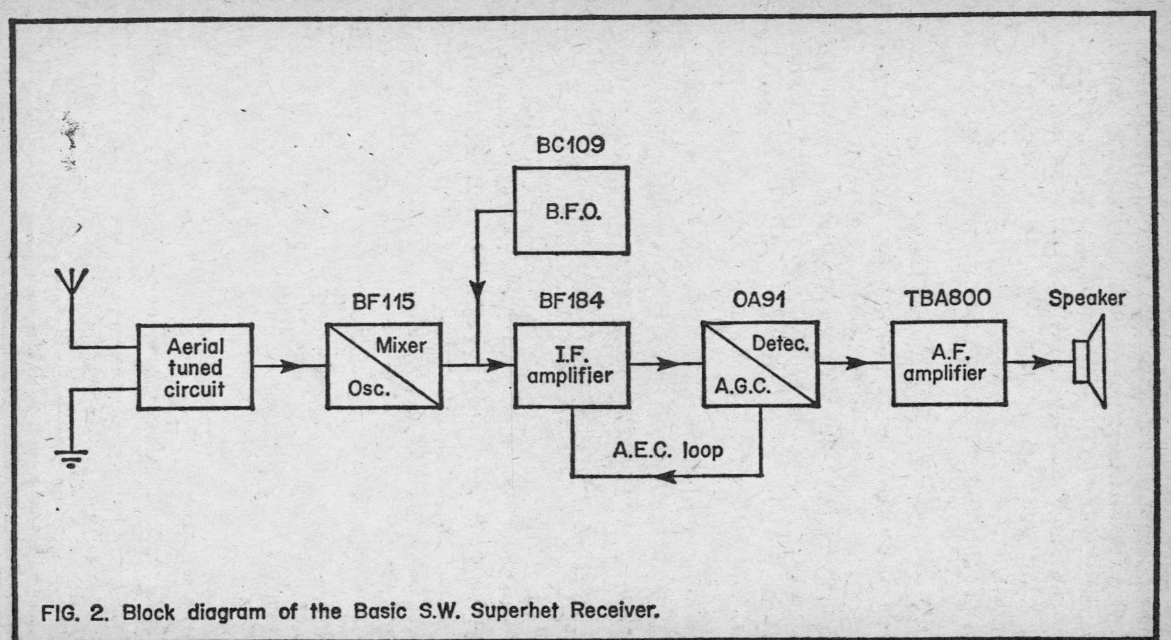

FIG. 2. Block diagram of the Basic S.W. Superhet Receiver.

This set obviously has its limitations, but it will nevertheless provide worldwide reception on both the amateur and commercial broadcast bands. Full constructional details will be provided, and it is well worth studying this section even if no attempt to construct the receiver is to be made, since the general method used for this set can be applied to the more complicated designs which are covered subsequently.

As mentioned earlier, the set has been designed so that it is simple to construct and it should not prove to be too difficult for anyone of limited experience. It provides a good and relatively inexpensive introduction to superhet designs, and it is strongly recommended that anyone who does not have any previous experience with superhet sets should tackle this design before attempting to construct any of the more advanced designs.

The Circuit

The block diagram of Figure 2 shows the various stages of the design, and it also gives details of the semiconductors employed in each stage. There is basically very little difference between this and the block diagram of Figure 1, but there are a few differences which will be discussed before proceeding further.

The first difference is that separate mixer and oscillator stages are not used, but instead a single stage employing one transistor is used to carry out both functions. This practice is not uncommon in simple superhet designs, and in fact many quite advanced receivers use precisely the same basic circuit that is utilized here.

Another difference is the addition of a B.F.O. stage, and this is necessary to permit the proper reception of C.W. (morse) and S.S.B. (single sideband) signals. With a T.R.F. design of course, the detector can be used to provide a sort of B.F.O. action by adjusting it to beyond the threshold of oscillation. Superhet receivers usually employ an ordinary diode detector, and so this is not possible, and a separate oscillator stage is required.

It is more convenient to inject the B.F.O. signal into the I.F. stages rather than at the aerial, since the I.F. stages operate at a single frequency, and it is therefore only necessary for the B.F.O. to operate at a single frequency. Using the other method it would be necessary to use an extra gang on the tuning capacitor so that the B.F.O. would always be at the input signal frequency.

The third and final difference is the addition of an A.G.C. circuit. A.G.C. stands for 'Automatic Gain Control', and sometimes the alternative term A.V.C. (Automatic Volume Control) is used.

There are two main reasons for incorporating A.G.C. in receivers. The primary one is that there are very wide variations in the levels of received signals, with the strongest signals being something like 10,000 times the amplitude of the weakest signals which can be resolved. Having to continuously alter the gain of the circuit manually so that weak signals could be received, and to prevent overloading on strong signals, would soon become rather tedious. The purpose of the A.G.C. circuitry is to automatically reduce the gain on strong signals so that a virtually constant output level is obtained without the need for manual adjustments. The second reason for including A.G.C. circuitry is that it makes station fading less apparent.

As will be seen from Figure 2, the set uses just five semiconductor devices (three silicon transistors, one germanium diode, and one I.C.), but the set nonetheless achieves a fairly high level of performance. The I.C. audio stage provides an output power of about 500mW to an 8ohm speaker, and it can be used with any speaker impedance of 8ohms or more. Note however, that the higher the speaker impedance used, the lower the available output power. The set can also be used in conjunction with any normal type of earpiece or headphones.

The set has a frequency coverage of approximately 1.6 to 31MHZ in three bands, with the approximate frequency coverage of each band being as follows:–

Range 3T	1.6 to 5.3MHZ.
Range 4T	4.8 to 15MHZ.
Range 5T	10 to 31HMZ.

The range numbers quoted above are those used by the coil manufacturer (Denco). Plug in bandchanging is used for the sake of simplicity, although it would be possible to modify the unit for bandswitching. This is a general topic which will be covered later.

The Circuit

The circuit diagram of the receiver, less the Audio and B.F.O. stages, appears in Figure 3. Looking at this in fairly broad terms, Tr1 is the basis of the mixer/oscillator stage, Tr2 is the basis of the I.F. amplifier, and D1 is the detector/A.G.C. diode.

Looking at this in greater detail, the aerial signal is coupled to VR1 which is an input attenuator of the volume control type. This control is useful for the reception of S.S.B. signals, where the B.F.O. is easily swamped by strong signals. The output from the slider of VR1 is fed to the primary winding of the input transformer, T1.

VC1, VC2, and the main winding of T1 form the aerial tuned circuit, with VC1 being the main tuning control, and VC2 the aerial trimmer. The aerial trimmer is included in order that the aerial tuned circuit can be kept peaked at the correct frequency. This avoids the need for proper alignment of the aerial and oscillator tuned circuits in the conventional manner, and it probably gives slightly improved results since there should never be any misalignment. It does have the slight disadvantage that there is an extra control to adjust, but this is not really a major drawback, and this arrangement is becoming increasingly popular.

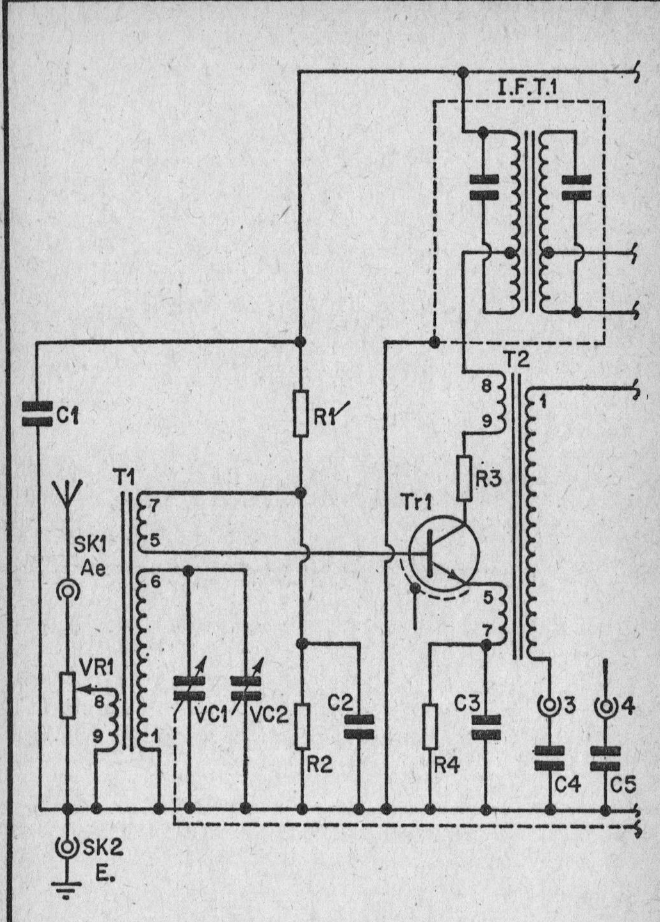

FIG. 3. The circuit diagram of the Basic Superhet Receiver. (Less audio and B.F.O. stages).

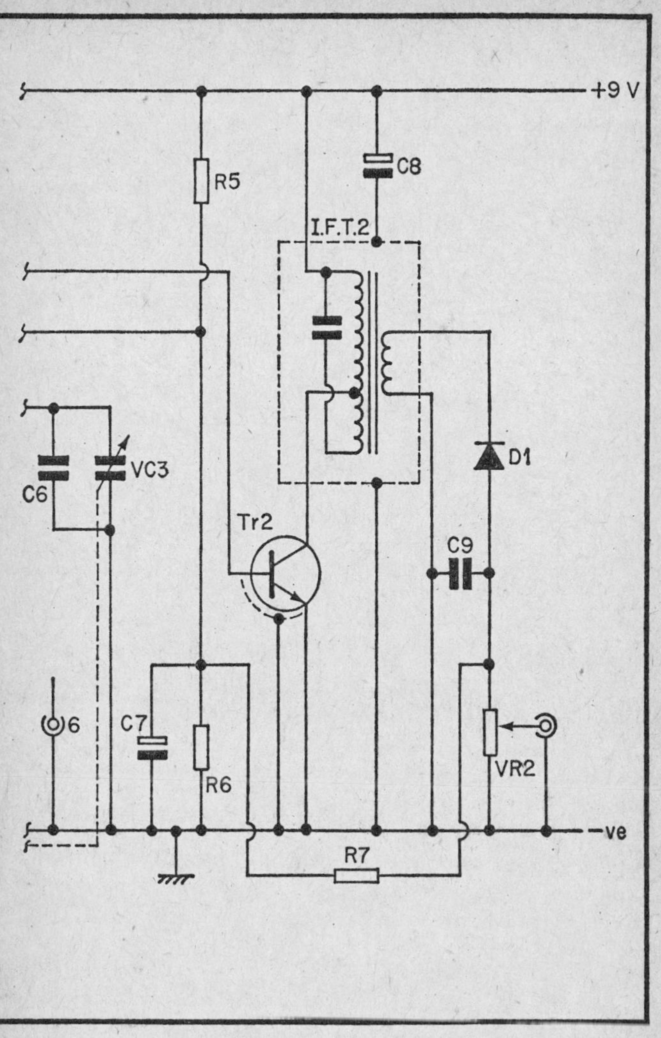
+9 V
R5
C8
I.F.T.2
VC3
C6
D1
Tr2
C9
6
C7
R6
VR2
-ve
R7

Tr1 is really used in two amplifying modes; as a common emitter amplifier as far as the aerial signal is concerned, and as a common base amplifier as far as the oscillator circuitry is concerned. The aerial signal is fed to the base of Tr1 via the low impedance coupling winding on T1. One end of this winding is earthed at R.F. by way of C2 while the other end couples direct to the base of Tr1.

R1 and R2 form a potential divider which produce the bias voltage for Tr1. This voltage is fed to Tr1 base through the coupling winding on T1. R4 and C3 are the usual emitter bias resistor and bypass capacitor respectively. The amplified aerial signals appear in the collector circuit of Tr1.

T2 is the oscillator coil, and the main winding of this, together with C6 and VC3, forms the oscillator tuned circuit. The two tuning capacitors (VC1 and VC3) are ganged so that they are operated simultaneously by the tuning knob.

Padder Capacitors

C4 and C5 are padder capacitors, and these, in effect, alter the value of the oscillator tuning capacitor to the correct level. As was mentioned earlier, the oscillator operates at a higher frequency than the aerial tuned circuit. However, both tuned circuits have to cover an identical range of frequencies. For instance, if the receiver covers a range of 10 to 20MHZ and has an I.F. of 470kHZ, the aerial tuned circuit tunes 10 to 20MHZ while the oscillator tunes 10.47 to 20.47MHZ. In other words they both cover a range of 10MHZ.

However, the oscillator operates at a higher band of frequencies than the aerial tuned circuit, and for this reason the oscillator tuning capacitance needs to be lower than the one for the aerial tuned circuit. Things are further complicated by the fact that at relatively low frequencies there is quite a wide difference between the input and oscillator frequencies, if one looks at the figures proportionately. For example, if a receiver is operating at 1MHZ and has a standard 470kHZ I.F., the oscillator frequency will be 1.47MHZ, which is almost 50% higher than the signal frequency.

At higher frequencies percentage difference between the input and oscillator frequencies is very much less. As a result, this receiver (and any other general coverage superhet of this general type) requires a different value of tuning capacitor for each range that is covered. On the L.F. bands the required oscillator tuning capacitance will be much smaller than that for the aerial tuned circuit, while on the H.F. bands there is no significance between the two values which are needed.

The padder capacitors are, in effect, added in series with the oscillator tuning capacitor, and their purpose is to reduce the capacitance swing of VC3 to the correct level. On Range 5 there is not a very great difference between the oscillator and aerial frequencies, and so in this case the earthy end of the oscillator tuned winding is connected to chassis directly.

The earthy end of the oscillator tuned winding connects to a different pin number on each of the three oscillator coils (which have a standard B9A valve type base). It connects to pin 3 of the Range 3 coil, pin 4 of the Range 4 coil, and pin 6 of the Range 5 coil. This is so that where plug in bandchanging is used, as it is here, by connecting the padder capacitors to the appropriate pins of the B9A coilholder for the oscillator coils, the correct padder is automatically selected when an oscillator coil is plugged in. In the case of the Range 5 coil of course, there is no padder capacitor, and pin 6 is simply earthed.

As an oscillator, Tr1 operates as a simple inductive feedback common base circuit, with the collector to emitter feedback being provided by way of the two smaller windings on T2. At first sight it may appear as though the base of Tr1 is not in fact grounded, but as far as the oscillator signals are concerned, Tr1 base is in fact earthed through the coupling winding of T1 and through C2.

The basic way in which this circuit provides the required mixing action is quite simple. The oscillations of Tr1 result in variations in collector current in sympathy with this signal. The gain of Tr1 varies with changes in collector current, and broadly speaking the gain rises and falls as the collector current rises and falls. Thus the oscillations modulate the aerial signal and the required mixing or heterodyning is produced, with the output appearing in the collector circuit of Tr1.

I.F. Stages

I.F.T.1 is a double tuned I.F. transformer, and this has its primary winding connected in the collector circuit of Tr1. It therefore receives the 470kHZ I.F. signal and passes it on to the I.F. amplifier stage.

Tr2 is the only active device which is used in the I.F. stage, and this is used as a high gain common emitter amplifier. It is biased by R5 and R6, and it receives this bias via part of the secondary winding of I.F.T.1. C7 earths the lower end of I.F.T.1 secondary.

Some readers may be wondering why tappings are used on the tuned windings of the I.F.T.s. The reason for this is simply that the impedance involved with bipolar transistors are fairly low, and this would result in heavy damping of the high impedance tuned windings of the I.F.T.s if the connections were to be made across the complete windings. This would also be rather inefficient. Therefore the low impedance tappings on the I.F.T.s have to be used in order to obtain high levels of gain and selectivity.

The output of Tr2 is developed across I.F.T.2 primary, and the untuned secondary winding of this I.F.T. feeds an ordinary diode detector which uses D1 and C9. VR2 is the volume control.

A.G.C. Action

Apart from the demodulated audio signal, a D.C. signal is produced across VR2. This is generated from the smoothed R.F. half cycles which appear across VR2, and the amplitude of this D.C. potential is proportionate to the amplitude of the received signal. The polarity of D1 is such that this voltage is negative with respect to chassis. This is important, and unlike some simple receivers where the polarity of the detector diode is irrelevant, it is essential that D1 is connected with the polarity shown.

The top of VR2 is connected to the potential divider which is used to bias Tr2, by way of R7. As a result of this, when a strong signal is received the voltage at the top of VR2 goes negative of chassis. As this voltage goes below chassis potential it tends to reduce the voltage at the junction of R5 and R6. This is due to the coupling between these two points via R7.

Thus the stronger the input signal the lower the bias voltage which is fed to Tr2. A lower bias voltage means reduced collector current, and hence also reduced gain. Therefore the stronger the input signal the lower the I.F. gain. Therefore a simple A.G.C. action is provided with full I.F. gain being produced on weak signals, and very little gain being produced at all on strong signals. Wide variations in the input signal amplitude produce only minor variations in the audio output level.

Audio Stages

An I.C. audio power amplifier forms the basis of the audio stages of the receiver, and this provides a high quality output which contains very little noise or distortion.

It is important that no R.F. signals should be allowed to break through into the audio amplifier since this would almost certainly result in instability. A low pass filter is used at the input

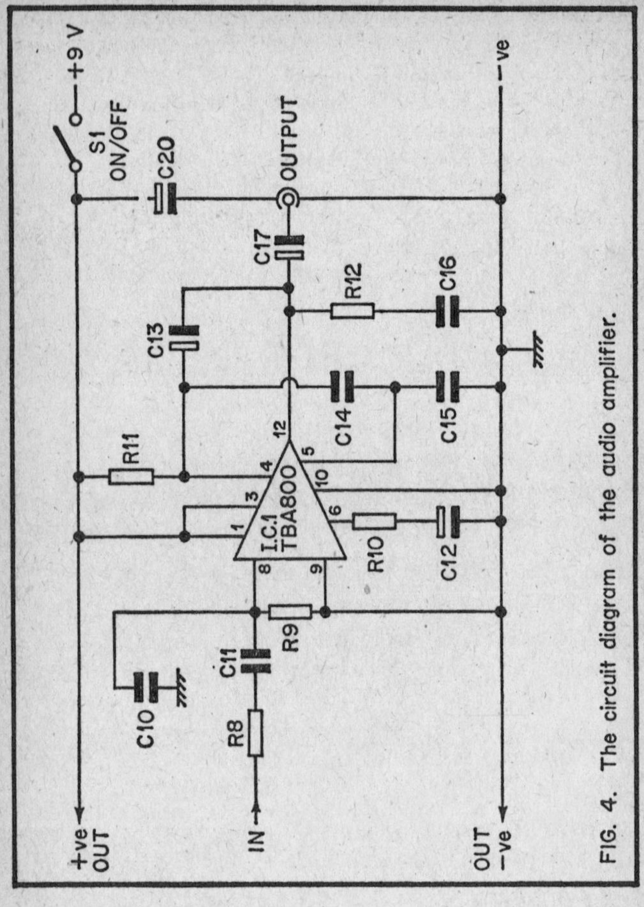

FIG. 4. The circuit diagram of the audio amplifier.

of the amplifier in order to ensure that no R.F. breakthrough does occur. This filter is formed by R8 and C10 of Figure 4, which is the complete circuit diagram of the audio stages of the receiver. C11 provides D.C. blocking at the input.

Briefly looking at the functions of some of the other components, C14, C15, R12, C16, and C20 are all required in order to suppress various form of instability. R11 and C13 provide bootstrapping between the output and driver stages, and this increases the output drive of the circuit. R10 sets the voltage gain of the circuit, and C12 is included in series with it to provide D.C. blocking. R9 is used to bias the input transistors of the circuit. Finally, C17 is the output D.C. blocking capacitor.

S1 is the on/off switch for the entire receiver.

B.F.O.

The circuit diagram of the B.F.O. appears in Figure 5, and this is a simple Colpitts circuit. Tr3 is used as a common base amplifier and positive feedback is provided between its collector and emitter by way of C19.

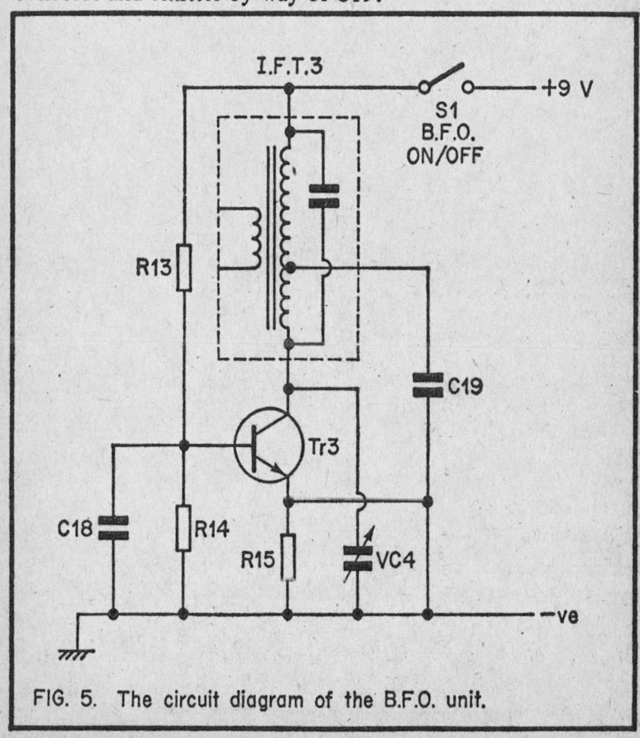

FIG. 5. The circuit diagram of the B.F.O. unit.

COMPONENTS (Figures 3, 4, and 5)

Resistors. All are miniature ¼ or 1/3 watt 5% types.

R1	15k	R9	68k
R2	12k	R10	56 ohms
R3	220 ohms (see text)	R11	150 ohms
R4	2.7k	R12	1.8 ohms
R5	56k	R13	39k
R6	15k	R14	33k
R7	10k	R15	1.8k
R8	12k		

VR1 1k lin. carbon.
VR2 5k log. carbon. with switch (S1).

Capacitors

C1 0.01mfd. ceramic.
C2 0.015mfd. plastic foil or ceramic.
C3 0.015mfd. plastic foil or ceramic.
C4 1100pf (see text).
C5 3000pf (see text).
C6 22pf polystyrene or mica.
C7 10mfd. 10v.w. electrolytic.
C8 100mfd. 10v.w. electrolytic.
C9 0.01mfd. type C280.
C10 5.6nf polystyrene.
C11 47nf type C280.
C12 100mfd. 10v.w. electrolytic.
C13 100mfd. 10v.w. electrolytic.
C14 330pf ceramic plate.
C15 2200pf plastic foil.
C16 0.1mfd. type C280.
C17 220mfd. 6.3v.w.
C18 0.015mfd. type C280.
C19 0.015mfd. type C280.
C20 330mfd. 10v.w. electrolytic.
VC1/3 Jackson 365 + 365pf air spaced type 02.
VC2 Jackson 50pf air spaced type C804.
VC4 Jackson 15pf air spaced type C804.

Inductors

T1	Denco Blue Aerial coils for transistor useage ranges 3T, 4T, and 5T.
T2	Denco Red Oscillator coils for transistor useage ranges 3T, 4T, and 5T.
I.F.T.1	Denco I.F.T.18/465kHZ.
I.F.T.2	Denco I.F.T.14/470kHZ.
I.F.T.3	Denco I.F.T.14/470kHZ.

Semiconductors

Tr1	BF115.
Tr2	BF184.
Tr3	BC109.
D1	OA91.
I.C.	TBA800.

Miscellaneous

18s.w.g. aluminium for chassis and panel.
Two B9A valveholders.
Plain 0.15in. matrix board and 0.1in. matrix stripboard.
3.5m.m. open construction jack socket and two wander sockets.
Five control knobs.
PP6 battery and connectors to suit.
Eagle 37m.m. 8 : 1 reduction tuning drive (type T501).
Rotary on/off switch (S2).
Wire, solder, hardware, etc.

VC4 enables the B.F.O. frequency to be adjusted slightly either side of the central intermediate frequency, so that upper and lower sideband signals can be satisfactorily resolved.

The output of the B.F.O. is very loosely coupled to the I.F. amplifier by means of a capacitive coupling. This coupling merely consists of a length of insulated wire which is taken from Tr3 collector and placed near the I.F. amplifier circuitry. It is not advisable to use a more positive coupling since a high level of B.F.O. injection would operate the A.G.C. circuitry and result in a reduction in sensitivity. Incidentally, it is because of this fairly low level of B.F.O. injection that the set is easily overloaded by strong S.S.B. signals, and this is why it is necessary to include the input attenuator control. Another reason for using a low level of B.F.O. injection is that using a higher level can often cause problems with microphony and instability.

Of course, the B.F.O. is only needed if the set is to be used for amateur bands reception where C.W. and S.S.B. are the main transmitting modes in current use. If the constructor is only interested in broadcast band reception, where A.M. is the only mode of transmission which is used, then all the B.F.O. circuitry should be omitted.

General Points

Finally, one or two minor points concerning the circuit will be considered. For example, the resistor (R3) in the collector circuit of Tr1 may appear to serve no useful function; just the opposite in fact. The reason for its inclusion is that at the high frequency end of each band the circuit will tend to be a little lively with instability manifesting itself in the form of whistles as the set is tuned across stations.

R3 is included to damp down this instability, and it will probably be necessary to experiment a little in order to find the optium value for this component. This aspect is fully dealt with later on.

In common with many R.F. transistors, Tr1 and Tr2 have four leadout wires, the additional one being a shield connection. This merely connects to the metal case of the device, and when a transistor having this extra leadout is used as an R.F. or I.F. amplifier it is usual to connect the shield to earth. This can aid good stability. When such a device is used as an oscillator there is obviously little point in connecting the shield leadout to chassis. This is why the shield connection of Tr1 is not earthed, while that of Tr2 is earthed.

The value of C7 may seem to be rather high for an earth return capacitor operating at 470kHZ. However, it also plays a secondary role in that together with R7 it acts as a lowpass filter. These two components thus allow the D.C. A.G.C. bias voltage to pass, but they block the audio signal present across VR2.

C1, C8, and C20 are the supply decoupling capacitors for the mixer, I.F. amplifier, and A.F. amplifier stages respectively. These are essential if feedback through the supply lines and consequent instability are to be avoided.

Chassis and Panel

The receiver is constructed using a home made chassis and front panel, and these are both constructed from 18 s.w.g. aluminium. Details of both the chassis and panel are provided in Figure 6.

Most of the drilling and cutting is quite straightforward. The three ¾in. (19m.m.) diameter cut outs are made using a chassis punch. When the two ¾in. diameter holes in the chassis have been made the positions of the two smaller holes for the coilholders are located using the coilholders themselves as templates. The coilholders are actually unskirted B9A valve-holders, and they must be high quality types. The coilholders are orientated so that pins 1 and 9 are roughly towards the front of the chassis. The smaller mounting holes are drilled

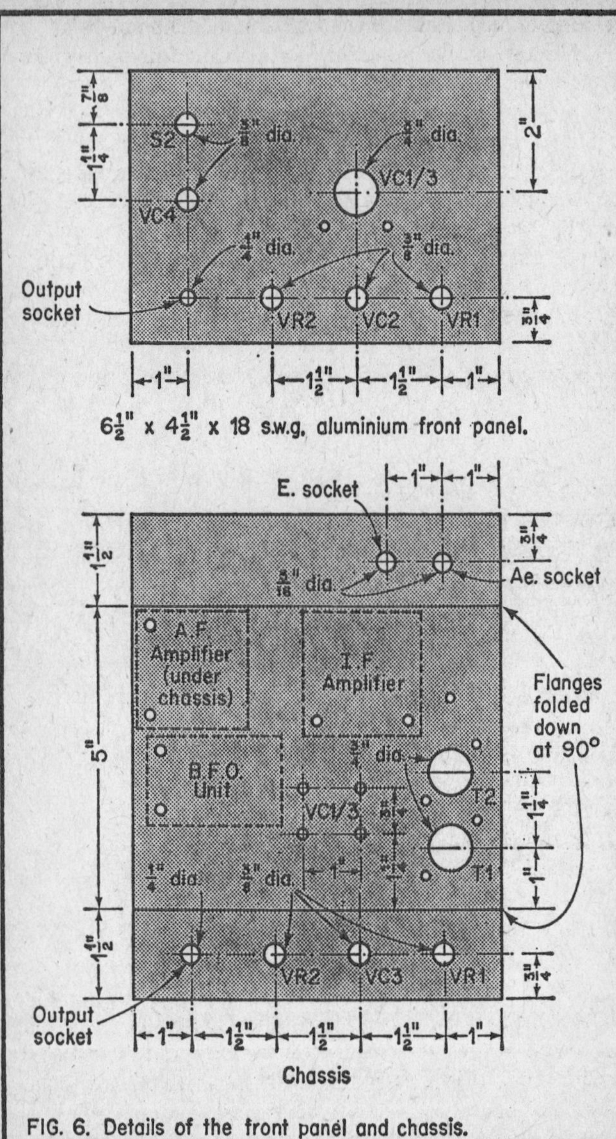

FIG. 6. Details of the front panel and chassis.

for 6BA (or M3) clearance using a 3.3m.m. diameter twist drill. The two smaller mounting holes for the tuning drive are not drilled at this stage.

There are four small holes in the chassis which serve no apparent purpose, and these are merely where insulated leads will eventually pass through the chassis. These should be about 3/16in. diameter and their precise positioning is not critical. It is advisable to fit them with small grommets, but this is not essential. The two mounting holes for VC1/3 are drilled for 4BA clearance using a 4m.m. diameter twist drill.

When all the drilling has been completed the two flanges of the chassis are bent down at right angles. Aluminium is fairly soft and pliable, and by using a little ingenuity it should be possible to make these two folds even if no special equipment for this purpose is available.

The chassis and panel are, in effect, bolted together by the mounting bushes and nuts of VR1, VR2, VC3, and the output socket. Then the two coilholders are mounted on the chassis, and a soldering is secured under one mounting nut of each holder (see Figure 7). The tuning capacitor is mounted by two short 4BA bolts which pass through the holes in the chassis and into threaded holes in the underside of the capacitor. It is important that these bolts should no more than fractionally penetrate through the base plate of the capacitor, as if this should be allowed to happen, the mounting screws could damage the fixed vanes of the capacitor. Either the mounting screws must be no more than about 5/16in. long, or washers must be used over them on the underside of the chassis in order to reduce their penetration to an acceptable level.

When the tuning capacitor has been mounted, the tuning drive can be placed in position and it can then be used as a template with which the positions of the two smaller mounting holes can be located. The chassis and panel assembly is then dismantled so that these two holes can be drilled to accept 6BA screws. It will also be necessary to carefully enlarge the two mounting holes in the tuning drive so that they will accept 6BA screws. The chassis and panel are then reassembled and the tuning drive is mounted using a couple of ½in. 6BA screws with nuts. The tuning drive is supplied with a short screw which fits into a threaded hole at the rear and towards the top of the drive. This mounting screw is not used here, and is discarded.

Before tightening the screw which clamps the output of the drive to the spindle of the tuning capacitor, adjust the drive to read '10' and adjust the tuning capacitor so that its two sets of vanes are fully meshed. The adjustment ranges of the two will then properly coincide.

Mixer Wiring

Point to point wiring is used in the construction of the mixer, and all this wiring is illustrated in Figure 7. This diagram also shows the other below chassis point to point wiring of the receiver.

It is not essential to use point to point wiring when constructing the input circuitry of a S.W. superhet receiver, and plain matrix S.R.B.P. panels, p.c.b.s., etc. can be used. However, it is usually quicker and easier to use point to point wiring, and it has the advantage long connecting wires in parts of the circuit which handle R.F. signals are easily avoided. It should perhaps be pointed out that many component leads are shown as being left quite long in Figure 7 in order to aid the clarity of the diagram. In practice all leads are kept as short and direct as possible.

Most of this wiring is quite simple and straightforward. When soldering in C2 and R2 it is probably easier if these two components are connected together before being soldered into position. The same is also true in the case of R4 and C3.

The two padder capacitors have rather unusual values, and some difficulty may be experienced in obtaining suitable

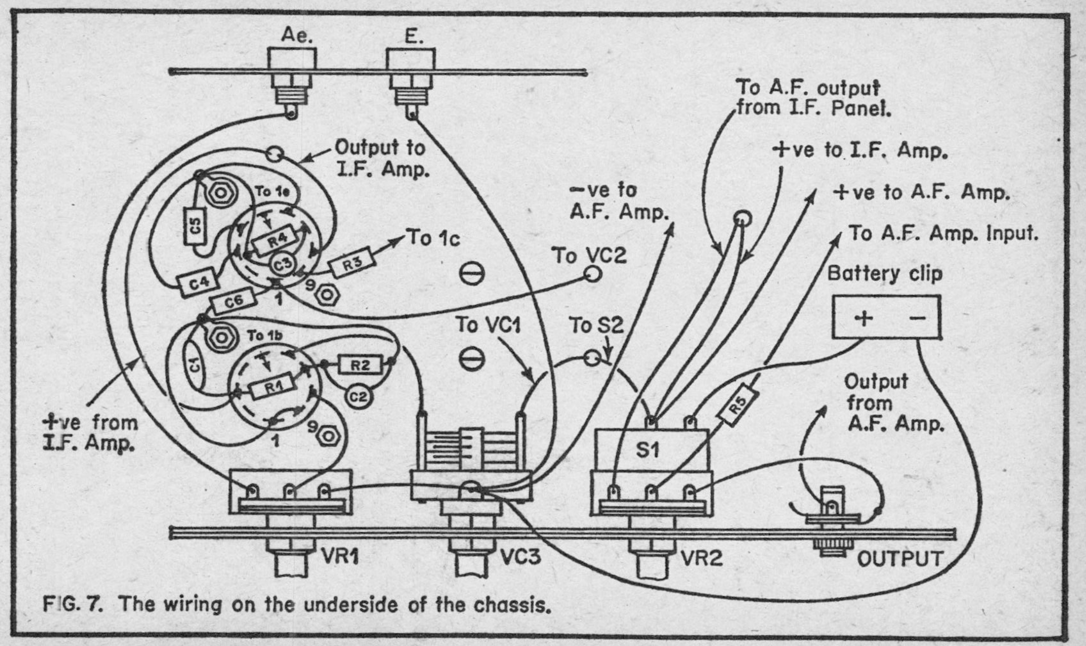

FIG. 7. The wiring on the underside of the chassis.

components. One of the larger mail order component firms could probably supply suitable types, or alternatively these values can be made up by adding two components in parallel. For example, 1100pf can be made up from a 100pf and a 1000pf capacitor wired in parallel. Similarly the value of 3000pf can be made up from two 1500pf components.

Whether these capacitors are made up from two parallel connected components or single capacitors are used, they must in either case be good quality low tolerance types such as polystyrene or silvered mica capacitors.

I.F. Amplifier Wiring

The I.F. amplifier, detector, and A.G.C. circuitry is constructed on a plain 0.15in matrix S.R.B.P. panel having 15 x 14 holes. Details of both the component layout and underside wiring of this panel are shown in Figure 8.

When a board of the appropriate size has been cut out the two 6BA or M3 clearance mounting holes are drilled out. It is also necessary to enlarge the holes into which the pins and mounting lugs of the two I.F.T.s fit. This can be done using a twist drill of about 2.5m.m. in diameter. I.F.T.1 has two tuning cores, one of which can only be adjusted from below. It is therefore necessary to enlarge the appropriate hole in the board so that it is possible to gain access to this core.

After the board has been prepared the various components are mounted and their lead out wires are bent flat against the underside of the panel. In the case of the I.F.T.s the mounting lugs are bent inwards at right angles on the underside of the panel, and this should firmly secure the I.F.T.s. Then the wiring on the underside of the panel is completed. In one or two places the component leads will probably be found to be too short, and it will be necessary to use approx. 22 s.w.g. tinned copper extension wires to bridge the gaps in the wiring.

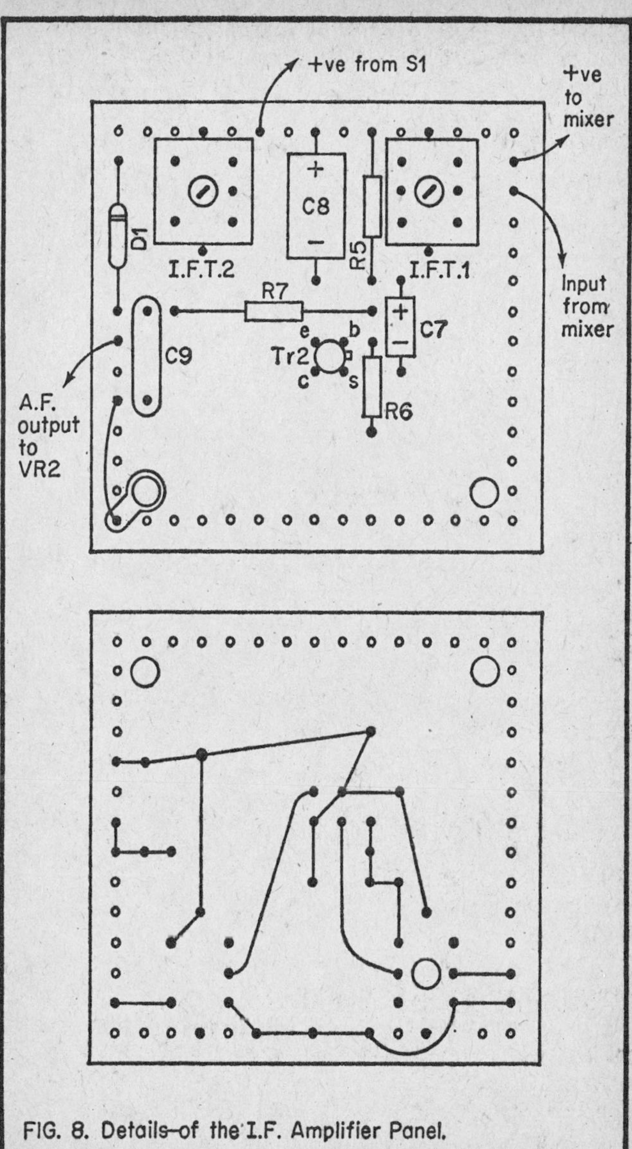

FIG. 8. Details of the I.F. Amplifier Panel.

The panel is then used as a template with which the positions of the mounting holes in the chassis are located. The approximate position in which the panel is mounted is shown in Figure 6. Before finally bolting the board in place, wire it up to the rest of the unit. Spacers about 10m.m. long are used over the mounting bolts, between the chassis and the board, so that the underside wiring is held clear of the metal chassis.

It is a good idea to drill a hole in the chassis beneath I.F.T.1 so that its core can be adjusted from below the chassis. Alternatively, when it comes to carrying out the I.F. alignment the panel can be dismounted, and a temporary earth connection to the board can be made (the normal earth connection is made via a short lead which connects to a soldertag mounted on one of the mounting bolts of the board).

Audio Stages

The audio amplifier is constructed on a 0.1in. matrix stripboard which has 25 holes by 18 copper strips. Details of this panel appear in Figure 9.

Start by cutting out a panel of the correct size and then drill out the two mounting holes. Next make the six breaks in the copper strips and then solder in all the components and link wires, with the I.C. being left until last.

The TBA800 I.C. is contained in a sort of 16 pin quad in line package, but where pins 4, 5, 12, and 13 would normally be, the device has a couple of heat tabs instead. Thus the I.C. actually only has 12 pins.

Normally the heat tabs would be soldered to an area of copper laminate on a p.c.b., with the copper then acting as a heatsink. However, here the device is used well below its maximum power rating of 5 watts, and so no heatsinking is required. The heat tabs should therefore be bent up out of the way, or even cut off altogether.

The completed panel is wired up to the rest of the unit before being mounted on the underside of the chassis in the approximate position shown in Figure 6. It is mounted in the same way as the I.F. amplifier panel.

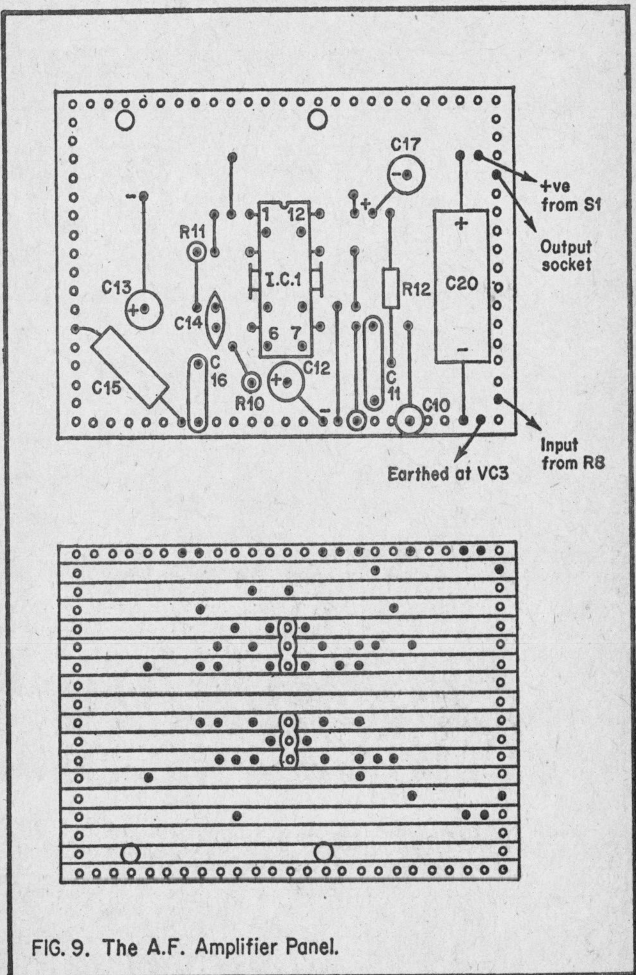

FIG. 9. The A.F. Amplifier Panel.

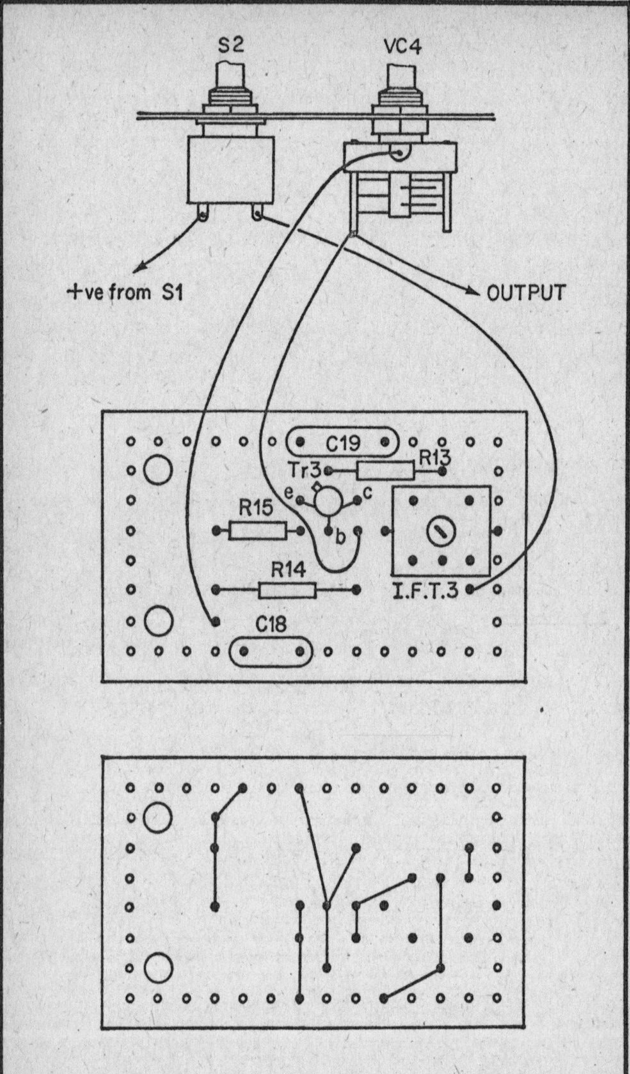

S2
VC4
+ve from S1
OUTPUT
C19
R13
Tr3
e
c
R15
b
R14
I.F.T.3
C18

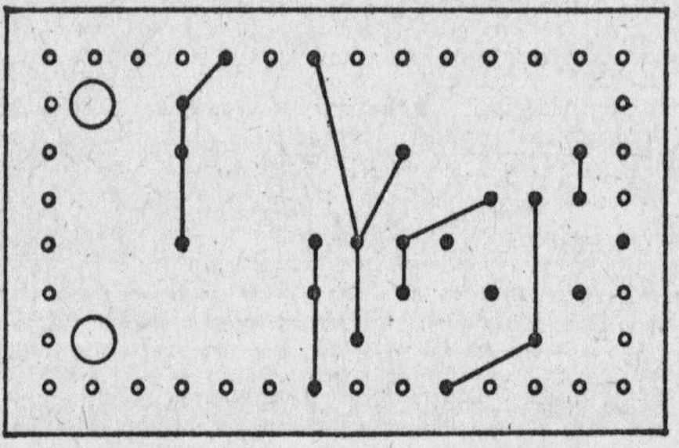

FIG. 10. Details of the B.F.O. Unit.

B.F.O. Unit

This is constructed on a plain 0.15in. matrix panel having 14 by 8 holes, and details of this panel together with details of the other B.F.O. wiring appear in Figure 10. The B.F.O. is built along very similar lines to the I.F. amplifier circuitry.

A lead about 150m.m. long carries the B.F.O. output to the I.F. amplifier, but as was stated earlier, no direct connection to the I.F. amplifier is made. A very loose coupling is all that is required, and it is merely necessary to place the B.F.O. output lead near the I.F. amplifier panel. It is advisable to tape this lead in place, rather than just leaving it to hang loose. Apart from anything else, this will provide a more stable B.F.O. signal.

Alignment

When initially testing the receiver it is best to use the Range 4T coils as this range will almost invariably provide numerous strong signals. The Blue aerial coil plugs into the front coil-holder and the Red oscillator coil plugs into the rear one. With an aerial and a speaker (or headphones) connected to the set, VR1 fully advanced and VR2 well advanced, it should be possible to tune in several stations.

For those who are unfamiliar with the Denco coil units it should perhaps be pointed out that as supplied they have their tuning cores virtually fully screwed down for packing purposes. In use the cores are unscrewed so that about 10m.m. of metal screwthread protrudes from the top of each coil. When some stations are received it should be possible to adjust the core of T1 to peak these stations.

Alignment of the set starts with the adjustment of the I.F.T. cores. Tune to a fairly weak but consistant station, and then adjust the I.F.T. cores to peak this station. Do not choose a strong station because the A.G.C. action of the receiver will make it difficult aurally to detect maximum signal strength. Use a proper trimming tool when adjusting the I.F.T. cores as a wedge shaped tool such as a small screwdriver blade could damage the cores. Tools not specifically designed for trimming applications also tend to have a detuning effect on the I.F.T.s as they are withdrawn from them. A suitable trimming tool for use with the specified I.F.T.s is the Denco type TT5.

Adjust the cores of I.F.T.2 first, then the upper core of I.F.T.1, and finally the lower core of I.F.T.1. Repeat this procedure a couple of times to ensure that the cores are all accurately peaked.

Note that very little adjustment of these cores should be required as the I.F.T.s are prealigned. It is also worth noting that although I.F.T.1 has a nominal operating frequency of 465kHZ while that of I.F.T.2 is 470kHZ, this difference is so small that in practice the two are perfectly compatible.

Next the B.F.O. is aligned, and at the outset this is switched on by operating S2 and then VC4 is set at about half maximum capacitance. Tune the set accurately to a station and then adjust the core of I.F.T.3 until a loud whistle is heard from the speaker or headphones. It should be possible to alter the pitch of this tone by adjusting the core of I.F.T.3, and this is given the setting which gives the lowest possible tone. Adjusting VC4 either side of its central position should produce a rise in the pitch of the whistle.

Do not be mislead by any fairly quiet whistling sounds which might be produced as the core of I.F.T.3 is adjusted, these can be caused by harmonics of the B.F.O. being picked up in the aerial circuitry of the set. They are easily distinguished from the main B.F.O. signal being received by the I.F. circuitry, because the latter produces a much stronger signal pick up. Also, the main B.F.O. signal will be present whatever frequency the set is tuned to, whereas harmonics will only be received at certain settings of the tuning control.

The R.F. alignment is very simple, and it is merely necessary to tune to a station with VC1/3 at about half maximum capacitance, set VC2 at half capacitance, and then adjust the core of T1 to peak the received signal. This procedure must then be repeated with the other two sets of coils in circuit.

It is quite possible that instability will occur at the high frequency end (VC1/3 vanes unmeshed) of one or more of the ranges. Unscrewing the core of T2 might help to eliminate this problem, but remember that if the core of T2 is adjusted it will then be necessary to readjust the core of T1. If this does not completely cure the instability it will be necessary to increase the value of R3 slightly. The correct value is found by means of trial and error, and is the lowest value which will clear the trouble. Do not make this resistor any higher in value than is really necessary, because this will unnecessarily reduce the level of performance obtained at the low frequency end of each band.

Using The Set

The receiver is quite easy to use, and in fact superhets are more simple to operate than T.R.F. designs. VC1/3 are the ordinary tuning control, and VR2/S1 are a conventional combined volume and on/off control. VR1 is a sort of R.F. gain control, and it is normally left at maximum (adjusted fully clockwise). It may sometimes happen that an extremely strong A.M. signal will overload the detector and it will then be necessary to turn back VR1 in order to obtain a proper audio output. This will only happen very infrequently though, and only if a very efficient aerial is used. VC2 is the aerial trimmer, and this is always adjusted to peak received signals for maximum signal strength.

For the reception of C.W. or S.S.B. signals the B.F.O. is switched in by means of S2. The setting of VC4 is not too important when C.W. signals are being received, but for the reception of S.S.B. it should ideally be offset from its central position.

For reception of upper sideband, which is used on the 20, 15, and 10 metre amateur bands, the vanes of VC4 should be almost fully meshed. They should be almost fully unmeshed for reception of lower sideband signals. Lower sideband is used on the 160, 80, and 40 metre amateur bands.

Strong S.S.B. signals will overload the set by swamping the B.F.O. signal, and this will result in a rather rough and distorted audio output. In extreme cases the audio signal will be virtually unintelligible. In order to resolve strong S.S.B. signals properly it will be necessary to turn back VR1 to a point where no overloading occurs.

CHAPTER TWO

MODULAR SUPERHET

In the previous chapter the superhet principle was described, together with very complete details of the circuit operation and construction of a practical receiver. A sort of modular form of construction was used with the circuit being divided up into four main sections (the mixer/oscillator, the I.F. amplifier/detector/A.G.C., the audio amplifier, and the B.F.O.). This modular approach was purposely adopted for two reasons.

Firstly it happens to be a very convenient form of construction, and secondly it leads up to this chapter where a number of different superhet building blocks will be described. The idea here is that the constructor chooses the various building blocks which best suit his or her needs, and then assembles these to produce a complete receiver.

This approach is perfectly feasible because superhet receivers tend to naturally break down into such building blocks. Provided one does not stray from standardised parameters (470kHZ I.F.s, usual stage input and output impedances, output levels, etc.), no problems with incompatibility between building blocks will arise. These building blocks have been designed so that they will fit together properly, without any problems of compatibility occuring. Such problems are unlikely to arrise anyway, and, for instance, the audio stages of virtually any superhet receiver could be substituted for those of the receiver described in the previous chapter with perfectly adequate results being obtained.

Of course,it would not be a good idea to mix valve and semiconductor circuitry, and all the circuits in the book employ semiconductors rather than valves. If one wants the ultimate in performance from the early stages of the set it is possible to make a good case in favour of using valves in these stages, with a beam deflection valve being used in the mixer stage. However, most people prefer the convenience of semiconductors, which are easier to use and more readily available than special types of valve.

It is probably true to say that there is not a great deal of difference between the performance obtained from the best valve and semiconductor devices anyway.

Image Rejection

The problem of image rejection was raised in the previous chapter, and it was pointed out that the basic superhet receiver which was described there had only a limited amount of image rejection at high frequencies. In more sophisticated receivers it is normal to take steps to increase the level of image rejection, and before going on to consider some practical circuits the methods of increasing image rejection will be discussed.

There are two basic ways of improving image rejection, either add more tuned circuits ahead of the mixer, or use a higher I.F. In previous eras the former was the method most often used in good quality communications receivers. These often had two R.F. stages with a tuned circuit at the input to each one, plus a tuned circuit at the input to the mixer. This provides some three tuned circuits ahead of the mixer, and this gives a level of R.F. selectivity which provides excellent image rejection.

Unfortunately there are drawbacks to this arrangement, one of which is the generous R.F. screening which is required in order to prevent instability. Another is that including the oscillator tuned circuit there are no less than four tuned circuits which must be maintained in good alignment if good results are to be procured from this arrangement.

Another problem, and perhaps the most major one, is that of cross-modulation which is a form of intermodulation. This distortion inevitably results in any amplifier since a perfectly linear amplifier has yet to be invented. This distortion can result in the modulation of one signal in the receivers R.F. passband being heard on the main signal which is being received, even though the two stations are well separated. It is from this that the term cross modulation is derived.

In a practical situation the results of cross modulation are never likely to be heard in this way, since the R.F. passband of a receiver is quite wide and the R.F. and mixer stages are likely to be handling dozens or even hundreds of stations. Cross modulation is much more likely to manifest itself in the form of a general increase in the background noise level. This is due to the fact that most S.W. signals consist of a complex and changing range of frequency components, and as the components of each signal react with the components of all the other signals in the receiver's R.F. passband, a vast range of new signals are produced. Those that are converted by the mixer to frequencies which lie within the sets I.F. passband are heard as a jumble of noise at the output.

The reason that cross modulation is such an important factor is simply that no matter how sensitive and selective the I.F. stages of a set are made, it will not be possible to copy weak signals if they are drowned in noise resulting from cross modulation. One might think that the old system of having three tuned circuits ahead of the mixer would give a good cross modulation performance by reducing the R.F. passband, and therefore reducing the number of signals handled by the input stages of the receiver.

In practice this is not the way things work out, because the R.F. bandwidth of the set will still be fairly wide even if three tuned circuits are used, and in order to eliminate problems with cross modulation noise in this way it is necessary to have an R.F. bandwidth which is nearly as narrow as the I.F. bandwidth. This is not really a practical proposition in a receiver which is to cover more than a single frequency!

Having two R.F. stages actively encourages cross modulation because there is inevitably a degree of distortion in each of these amplifiers. Furthermore, the effect of these stages is to increase the signal levels encountered in the input stages of the set, which results in higher levels of distortion. This is particularly so by the time signals reach the mixer, as any really strong signals will have been boosted to a level which will drive the mixer outside its linear operating range. Severe cross modulation will then occur, making it impossible to copy weak stations.

The modern approach to minimising the problem of cross modulation is to use no R.F. stage, and a high frequency I.F. The two I.F.s most often used are 5.5MHZ and 9MHZ. These produce an image signal which is 11MHZ and 18MHZ respectively away from the main response. Even a single tuned circuit ahead of the mixer will then provide good image rejection right up to the higher frequency limit of the S.W. frequency spectrum.

However, neither home constructed or commercially produced receivers of this type are frequently encountered in amateur circles. This is due to difficulties which exist with this arrangement, the most major one being that of obtaining good selectivity. I.F. transformers incorporate ordinary L – C tuned circuits, and at frequencies in the region of 5.5 to 9MHZ L – C circuits simply would not produce the high degree of selectivity that a good communications receiver must possess. In order to obtain the required level of selectivity from a high I.F. it is necessary to use a crystal filter, and these tend to be rather expensive if purchased ready made. A good crystal filter can cost more than the remainder of the parts for a receiver! It is possible to construct ones own filter, but obtaining suitable parts is likely to be something of a problem. Aligning the completed unit can also cause problems.

An alternative to using a crystal I.F. filter is to use the double conversion approach. Here the output from the first I.F. amplifier is fed to a second mixer where the signal is heterodyned to a conventional 470kHZ I.F. Thus the high first I.F. provides good image rejection while the low second I.F. provides good selectivity.

This method is probably more common than the single conversion high I.F. one, but neither are as popular as the single conversion low I.F. type of receiver. This is the type of set that we will be concerned with in this chapter.

A design of this type has to be something of a compromise, but this is not to say that a very high level of performance cannot be attained. The building blocks which are described in the following pages can be assembled to produce an excellent S.W. set. Whichever of the front ends the constructor chooses (there are four alternatives) there will be two tuned circuits ahead of the mixer. The coils used in the front end circuits are modern ferrite cored types which have quite high Q values. They therefore provide a fairly narrow R.F. passband, and thus also a quite high degree of image rejection even at frequencies in the 20 to 30MHZ region.

Three of the front ends which are described are active units, and can provide a fairly high level of gain (20dB or more at most frequencies). All incorporate R.F. gain controls though, and the gain can be reduced to less than unity so that overloading of the mixer can be avoided. The remaining front end circuit is a passive type, and it does not therefore introduce any cross modulation directly, or by increasing the signal amplitude fed to the mixer.

When using any S.W. superhet it is worth bearing in mind that obtaining the best possible aerial signal and using maximum R.F. gain does not necessarily provide the best results. For example, some time ago the author had a quite conventional valved communications receiver which invariably failed to produce any stations at all on the 40 metre amateur band after dark. This band is noted for poor reception after dark due to its very close proximity to the 41 metre broadcast band, with many broadcast stations encroaching into the amateur band allocation. The very strong broadcast stations tend to cause severe cross modulation which obscures the comparatively weak amateur stations.

The rather unlikely cure to the problem was to disconnect the aerial and then place the aerial plug close to its socket. This provided only a very loose coupling between the aerial and the receiver, but with the A.F. gain well advanced it was still possible to copy many stations, including some amateur ones. In fact, stations as far away as South America were copied.

What was happening here is quite simple; although the strengths at which stations were received was being greatly diminished, the cross modulation was being reduced by an even greater amount. Therefore, stations that were originally drowned in the cross modulation noise were brought above this noise level by attenuating the aerial signal. This feature is something that should be kept in mind when using any S.W. superhet receiver.

Details of the various building blocks will now be given. Constructional information will not be provided since it is assumed that anyone undertaking construction of these circuits will have reached the stage where they can build practical units from a circuit diagram. The following chapter will describe some additional circuitry which can be used in conjunction with certain of these building blocks to provide a double conversion receiver. The final chapter will deal with add on circuits such as an 'S' Meter, A Multiplier, etc.

R.F. Amplifiers

The circuit diagram of an R.F. amplifier which employs a bipolar transistor as its active element is shown in Figure 11.

Operation of the circuit is quite simple with Tr1 being used in the common emitter amplifying mode. The potential divider formed by R1 and R2 provides the base bias voltage for Tr1, while R3 and VR1 form the emitter bias resistance. C3 is the

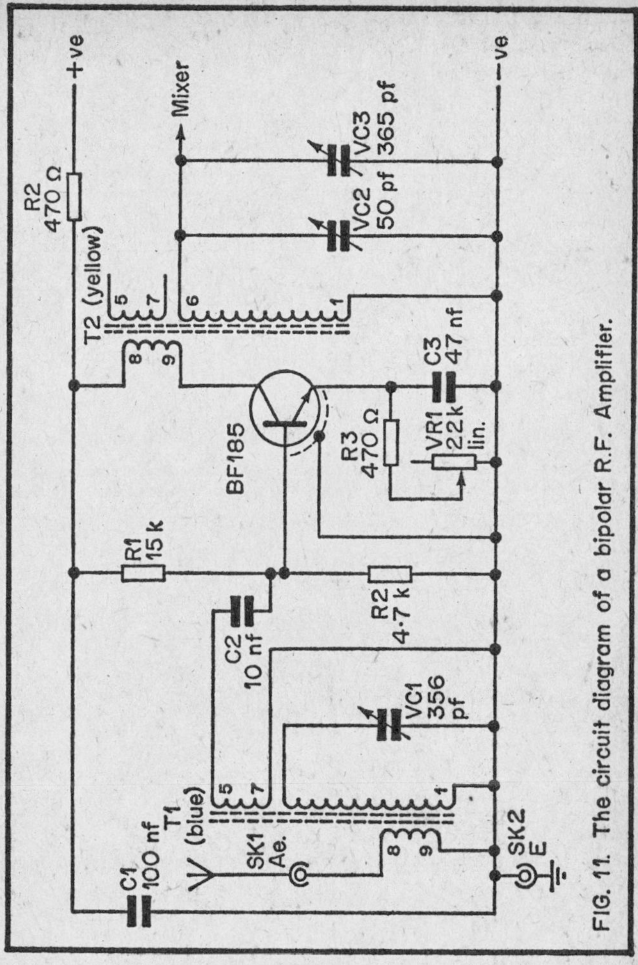

FIG. 11. The circuit diagram of a bipolar R.F. Amplifier.

usual emitter bypass capacitor. VR1 enables the operating current of the stage to be varied, and hence it also controls the gain of the stage. Gain is at maximum with VR1 set for minimum resistance. R3 is a current limiting resistor.

The base of Tr1 is fed from the low impedance coupling winding on T1 via D.C. blocking capacitor C2. VC1 is the tuning capacitor for the R.F. stage. The output of the R.F. amplifier is coupled to the mixer stage by way of R.F. transformer T2. VC3 is the tuning capacitor for the mixer tuned circuit and VC2 is a panel trimmer control which enables this tuned circuit to be kept in proper alignment.

It is not recommended that VC1 should be ganged with VC3 (which is ganged with the oscillator tuning capacitor of course) because this makes it necessary to have the R.F. amplifier circuitry and the mixer/oscillator circuitry in close proximity to one another. If this is done it becomes absolutely essential to screen off the R.F. amplifier circuitry in order to obtain good stability. This is not as easy to accomplish as one might imagine.

In the authors opinion it is far better to have VC1 as a separate component with the R.F. circuitry physically well spaced from the mixer/oscillator circuitry. Screening then becomes unnecessary. Note that T2, VC2, and VC3 are only shown here to illustrate how the R.F. amplifier is connected to the mixer. These are actually part of the mixer circuit, and are positioned close to the rest of the mixer circuitry. These last two points are also relevant to the next two front end circuits which are described.

T1 is a Denco Blue aerial coil for transistor useage, and T2 is a Yellow R.F. coil for transistor useage. The following front end circuits all employ the same two coil types.

Jugfet Amplifier

The circuit diagram of an R.F. amplifier using a Jugfet (junction gate field effect transistor) is shown in Figure 12. This has a similar performance to the bipolar R.F. amplifier circuit, but it provides a slightly lower noise level and has a better cross modulation performance.

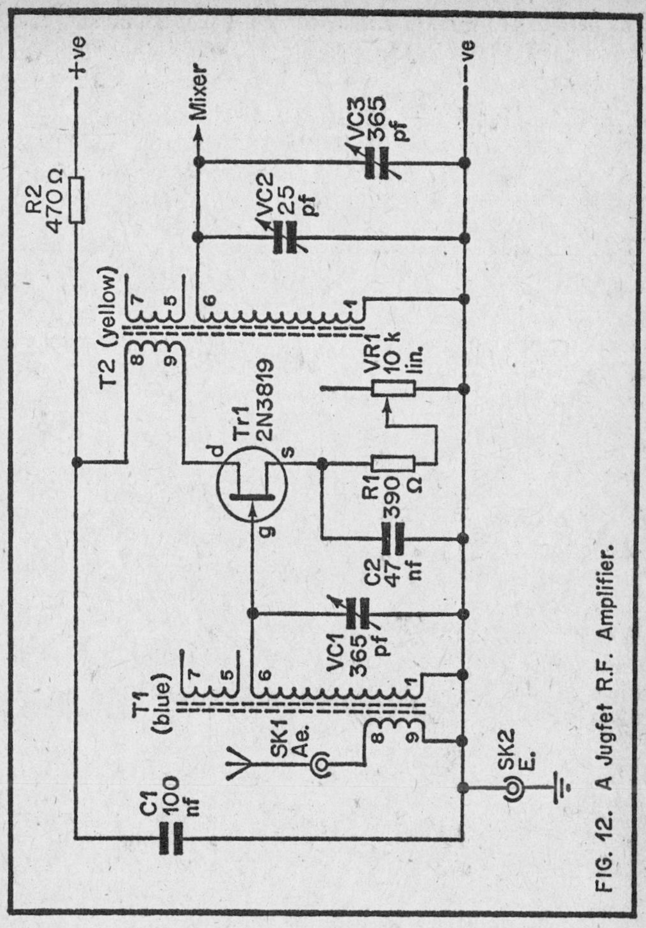

FIG. 12. A Jugfet R.F. Amplifier.

The gate terminal of Tr1 is bias by being tied to earth through the tuned winding of T1. There is no need to use the low impedance coupling winding of T1 in this circuit since f.e.t.s have extremely high input impedances, and the tuned circuit can therefore be coupled direct to the gate of Tr1.

R1 and VR1 form the source bias resistance for Tr1, and C2 is the source bypass capacitor. VR1 enables the standing current through the stage to be varied, and it thus controls the gain of VR1. As was the case with the previous circuit, gain is at a maximum with VR1 adjusted for minimum resistance.

R2 and C1 are a supply decoupling network.

MOSFET Amplifier

MOSFETs (Metal Oxide Semiconductor Field Effect Transistors) are somewhat more expensive than most other types of transistor, but this additional expense is easily justified by their superior performance when compared with bipolar and Jugfet devices in R.F. applications. The circuit diagram of a MOSFET R.F. amplifier is shown in Figure 13. This provides a slightly higher level of gain than the previous two designs, and it has a lower noise level.

Perhaps more importantly, it has a much better cross modulation performance, and MOSFETs are able to handle quite high signal amplitudes without producing high levels of distortion.

As was the case with the Jugfet R.F. amplifier, the high input impedance of the transistor enables the aerial tuned circuit to be directly coupled to the input gate. Tr1 is a dual gate MOSFET, and the input signal is coupled to the g1 terminal. The voltage gain at this terminal is determined by the g2 potential of the device, and VR1 can be used to vary this voltage from zero to about 2 volts. The gain of Tr1 is at a minimum when the gate 2 voltage is zero, and at maximum when it is at a level of about 2 volts. VR1 thus acts as the R.F. gain control. C1 filters out any stray R.F. signals which might otherwise be picked up at the g2 terminal, and result in instability or loss of performance. It also minimises any noise spikes which are produced at VR1 slider as this control is adjusted.

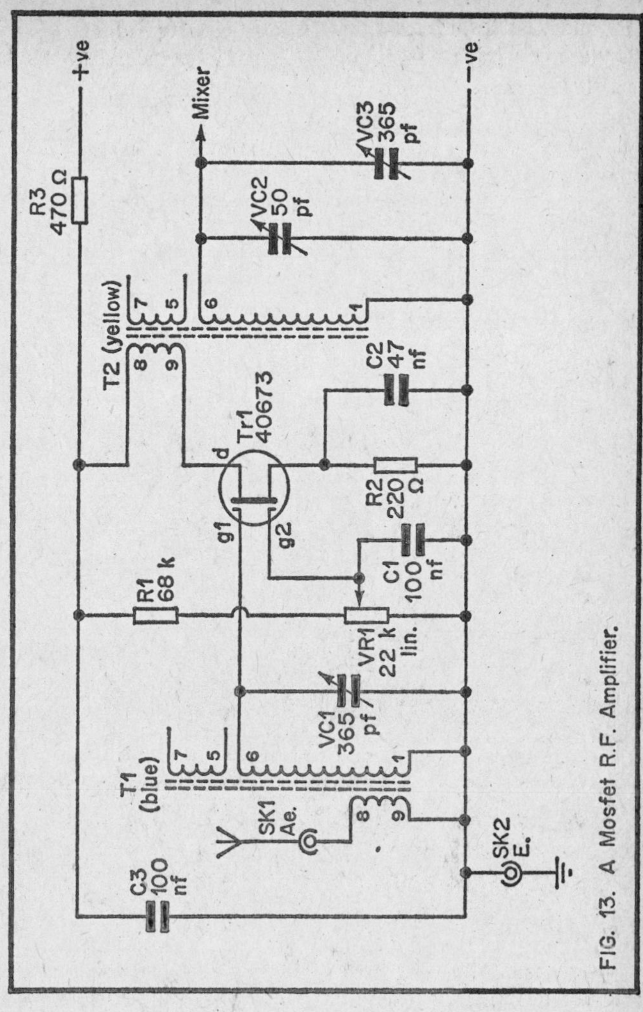

FIG. 13. A Mosfet R.F. Amplifier.

R2 and C2 are the source bias resistor and bypass capacitor respectively, and R3 and C3 are a supply decoupling network.

MOSFETs have a reputation for being easily damaged by static charges due to their extremely high input impedances (which are usually something in the region of 1,000,000 Meg. ohms). The 40673 transistor has integral protective diodes, and it is not therefore easily damaged in this way. It needs no more handling precautions than any normal semiconductor device.

Bandpass Input

It is quite possible to use more than one tuned circuit before the mixer without using any amplifying stages in this part of the circuit. Such an arrangement is usually termed a bandpass circuit, and the circuit diagram of such an arrangement appears in Figure 14.

Here the two R.F. transformers are merely coupled together by way of two low impedance windings. Of course, such an arrangement does not provide any gain, just the opposite in fact, and there is a significant loss through the input filter. However, this circuit produces no cross modulation of its own, and quite good results can be obtained using this arrangement.

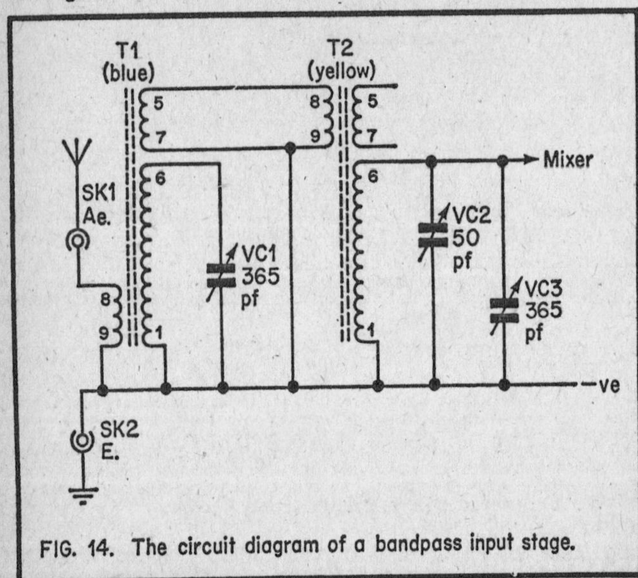

FIG. 14. The circuit diagram of a bandpass input stage.

Mixer/Osc. Stages

The mixer circuit which was used in the receiver which was described in the first chapter is simple and inexpensive, but as far as cross modulation performance is concerned it leaves a lot to be desired. As was stated in Chapter 1, the gain of the mixer transistor varies with changes in operating current. As the oscillator signal swings positive and negative the operating current decreases and increases, and as it does so it therefore modulates the input signal and the desired mixing action is produced.

The problem with this system is that the aerial signal also causes the operating current of the transistor to vary in precisely the same way as the oscillator signal does. This results in changes in the gain of the transistor which obviously causes distortion, including intermodulation distortion (of which cross modulation is a type). There is no simple way of eliminating this distortion, but even if there was there would be no point in doing so since if the distortion is removed, so is the mixing action. This type of mixer thus has an inherently high cross modulation level.

Dual gate MOSFETs make excellent R.F. amplifiers because, as was pointed out previously, they can handle high signal levels while producing only low distortion levels. it was also explained that the gain of the device was controlled by the voltage at the g2 terminal. Dual gate MOSFETs make excellent mixers, as by coupling the aerial signal to the g1 terminal and the oscillator signal to the g2 terminal, the oscillator will vary the gain of the device and thus modulate the aerial signal. A good mixing action is produced in this way.

Only low levels of cross modulation are generated by a dual gate MOSFET mixer because the gain of the device does not vary greatly with variations in the aerial signal amplitude, only with changes in the oscillator signal level.

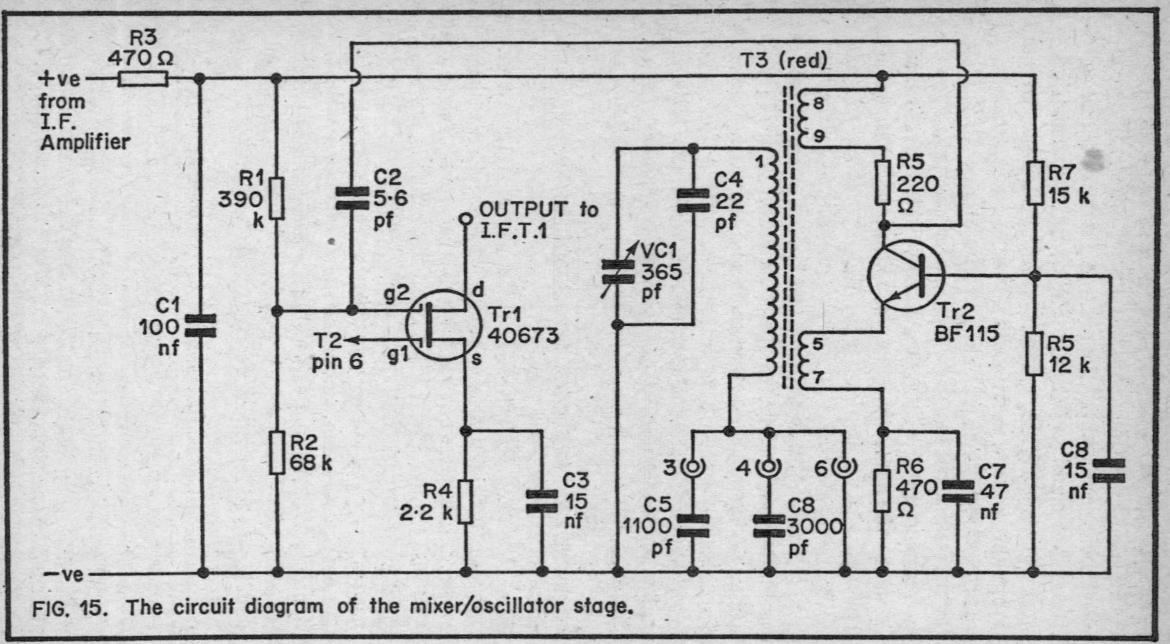

FIG. 15. The circuit diagram of the mixer/oscillator stage.

Since dual gate MOSFETs provide the basis of by far the best of simple mixer circuits, only this type of mixer will be considered here, there being little point in providing details of more complex designs of vastly inferior performance. The circuit diagram of a simple MOSFET mixer is shown in Figure 15, and this also shows the circuit of the recommended oscillator circuit.

The mixer tuned circuit is fed direct into the g1 terminal of Tr1, and the tuned winding on the mixer transformer (T2) provides the necessary D.C. path to earth for Tr1 g1. R1 and R2 form a simple potential divider which bias Tr1 g2 about 1 volt positive. The oscillator signal is coupled to the g2 terminal by way of C2, and by modulating the bias voltage this signal varies the gain of Tr1. R4 and C3 are the source bias resistance and bypass capacitor. The output of the mixer is taken from Tr1 drain, which connects to the first I.F. transformer of the I.F. amplifier. R3 and C1 are a supply decoupling network.

The oscillator circuit may look a little familiar, and this is because it is basically the same circuit that was used in the mixer/oxcillator stages of the receiver which was described in Chapter 1. As this stage is not coupled to the mixer transformer this time, the base bias network (R7 and R8) and earth return capacitor (C8) are connected direct to Tr2 base.

When used as a combined mixer/oscillator circuit it is necessary to select the circuit values to limit the amplitude of the oscillations, as otherwise instability is almost certain to occur. This is less of a problem when the circuit is used only as an oscillator, and the opposite was in fact found to be the case, with it being necessary to reduce the value of the emitter base resistor in order to increase the amplitude of the output signal to an acceptable level.

It is not quite as simple to construct this mixer/osc. circuit using a point to point wiring system as it was in the case of the mixer/osc. circuitry of the receiver described in Chapter 1, because more component anchor points are required here, and there are only the same number of spare coilholder pins which can be used. Note that where some coil windings are unused, no connections should be made to the relevant pins of the coilholder. Thus, for example, no connections should be made to pins 5 and 7 of the coilholder for T2 (Figures 12 and 13).

It is possible to solder some parts of the circuit together without using any anchor points, and so overcome this problem, but this is likely to give a rather flimsy assembly. This can result in stability of the oscillator frequency being affected by vibration, and it is not likely to provide adequate reliability. The method preferred by the author is to use point to point wiring with miniature tagstrips being mounted near the coilholders. The purpose of these tagstrips is to provide additional anchor points.

Alternative Oscillator

The oscillator circuit shown in Figure 15 is probably more stable (with regard to variations in frequency) than most people realise. It is unlikely that stability could be greatly increased without resorting to highly sophisticated circuitry, and this is really a complete subject in itself. Such things as frequency synthesisers are certainly beyond the scope of this book, and few constructors bother to go to such lengths anyway.

One simple way of improving the stability of the circuit of Figure 15 is to connect a 7.5 volt zener diode in parallel with C1 (cathode to the negative supply). This will substantially reduce any variations in the output frequency due to changes in the supply voltage (due to battery ageing etc.). This modification will similarly aid the stability of the alternative oscillator circuit which appears in Figure 16.

This employs a Jugfet in a simple common source inductive feedback oscillator. Only two of the windings on the oscillator

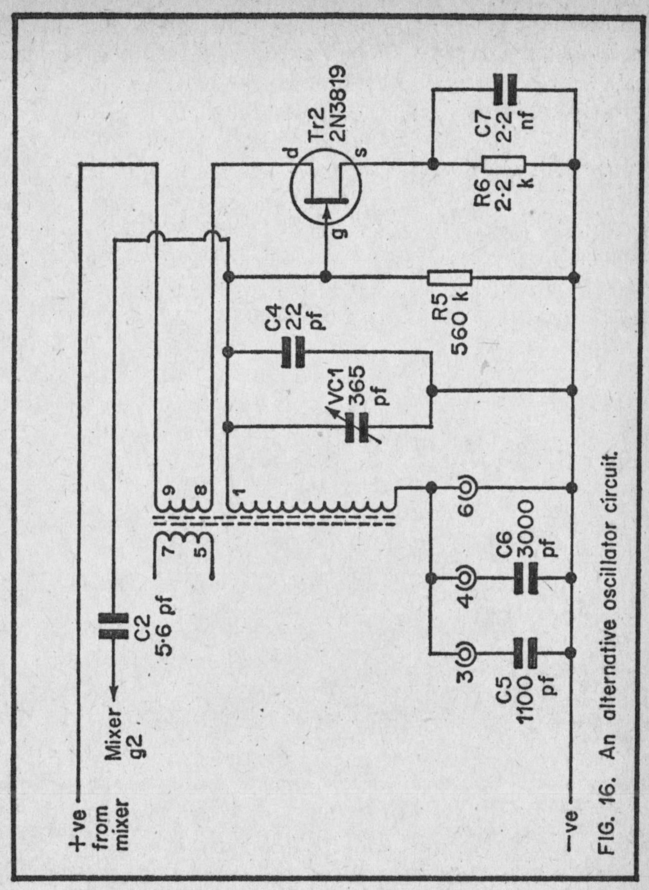

FIG. 16. An alternative oscillator circuit.

transformer are used in this circuit. When the range 3 and 4 coils are in circuit the tuned winding of the coil does not provide a D.C. path to earth for Tr2 gate due to the presence of the padder capacitors (C5 and C6). R5 has therefore been included to provide this D.C. path. The value of this resistor is more critical than one might think, and if it is made too low in value its shunting effect on the tuned circuit reduces the amplitude of the output signal, or oscillation ceases all together. On the other hand, if it is made too high in value it is not able to drain away the negative charge which builds up on Tr2 gate. As a result, this charge takes Tr2 gate about a volt or so negative of the earth rail, and oscillation is then halted as the transistor is cut off.

This circuit would appear to be slightly more stable than the one of Figure 15, but it does have a slight disadvantage. This is that it is difficult to obtain a sufficiently high signal amplitude from the circuit. The signal level of Tr2 drain (the obvious signal take off point) is far too low, and it is necessary to take the output from the tuned circuit instead.

On ranges 3 and 4 the circuit has a performance which is very much the same as that of the circuit of Figure 15 with regard to sensitivity, but on range 5 results would appear to be slightly inferior. The reason for this is probably that on range 5 the oscillator operates on the second harmonic and not on the fundamental. In other words the oscillator is actually operating at half the required frequency, and it is the second harmonic signal (which is at double the frequency of the fundamental) which is used as the oscillator signal. Because the f.e.t. circuit oscillates less violently than the bipolar one, it does not provide as much harmonic output, although the amplitudes of the fundamentals are very much the same.

I.F. Amplifier

If a simple I.F. amplifier is all that is required, the one described in Chapter 1 can be used. It is, however, more normal to use a two stage I.F. amplifier, and the circuit diagram of one of these appears in Figure 17. The fact that this circuit uses two transistors does not mean that it has something like one hundred times the gain (the gain that can be attained by a single 470kHZ I.F. amplifier) of the single transistor circuit described earlier. Even when using a very carefully and well planned layout, the gain that can be obtained is limited by stay feedback over the circuit. If the gain

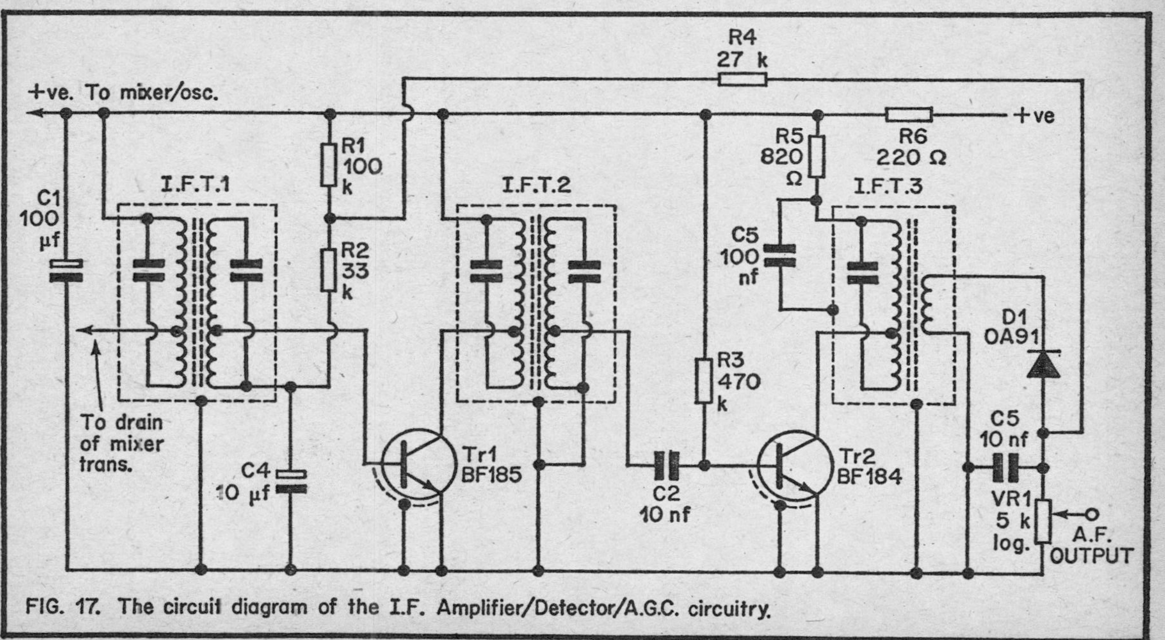

FIG. 17. The circuit diagram of the I.F. Amplifier/Detector/A.G.C. circuitry.

is made too high the level of stray pick up between the input and output of the circuit, will be sufficient to cause a high background noise level, or could even result in the I.F. stages breaking into oscillation. This means that it is necessary to use the two amplifying stages in the circuit of Figure 17 at something less than maximum obtainable gain in order to maintain stability. On the other hand, the single stage of the circuit of Figure 3 can be used at just about maximum gain. Thus there is only a limited, but very worthwhile increase in performance provided by using an additional I.F. stage.

Apart from increased gain, a second advantage of the two stage circuit is that it has an extra double tuned I.F. transformer, and this gives a significant increase in selectivity. I.F.T.1 and I.F.T.2 are both Denco I.F.T.18/465kHZ units, and I.F.T.3 is a Denco I.F.T.14/470kHZ type.

The circuit consists of two common emitter amplifier stages and diode detector D1. There is an A.G.C. loop via R4. Basically this circuit operates in much the same manner as the one described in the previous chapter, but it has another stage of amplification in the form of Tr2.

Although the circuit will perform very well with the values shown, it is possible to optimise gain by reducing the value of R3 slightly. The lower in value this component is made, the higher the gain. However, as mentioned earlier, the gain cannot be increased indefinately, because stray feedback will cause either excessive background noise or actual oscillation. R3 can therefore only be decreased in value to a point where neither of these are apparent, and no further.

A.F. Amplifiers

The I.F. amplifier circuit of Figure 4 works extremely well in this receiver. Another good I.C. audio amplifier circuit is shown in Figure 18, and this uses the popular LM38ON I.C.

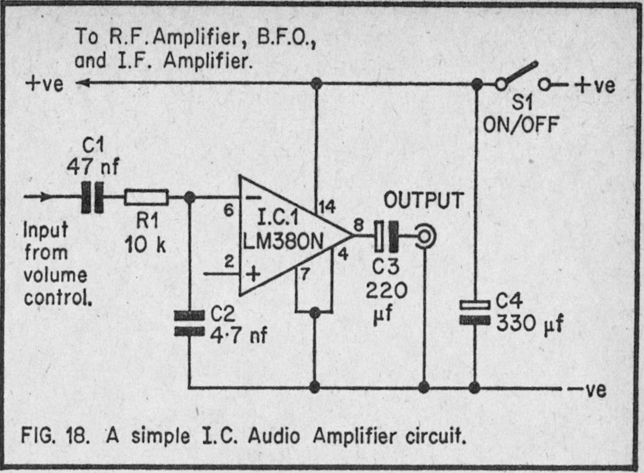

FIG. 18. A simple I.C. Audio Amplifier circuit.

This device requires very little in the way of discrete components, and it will work in some applications using only two discrete components. These are the input and output D.C. blocking capacitors, which are C1 and C3 in Figure 18. In this particular application it is necessary to employ three other discrete components.

In order to ensure stable operation of the device it is necessary to add a lowpass filter at the input so that no R.F. signal is fed into the I.C. R1 and C2 form this filter. They also restrict the upper frequency response of the circuit somewhat, and this gives a slightly improved signal to noise ratio. C4 is a supply decoupling capacitor, and this is required in order to prevent low frequency instability, or motor boating as it is sometimes called. S1 is the on/off switch for the entire receiver.

This circuit will provide about 500mW. into an 8 ohm impedance speaker, and it can be used with a higher impedance load if a reduction in maximum output power can be tolerated. It can also be used to drive virtually any type of headphones.

The LM380N incorporates short circuit protection, and also has thermal shut down protection circuitry, although the latter is of little importance in this low power application. Incidentally, the TBA800 I.C. which is employed in the A.F. amplifier described previously has neither form of protection circuitry. However, it is used well within its ratings in the circuit of Figure 4, and it is highly unlikely that either short circuiting the output or overdriving the circuit would destroy the I.C.

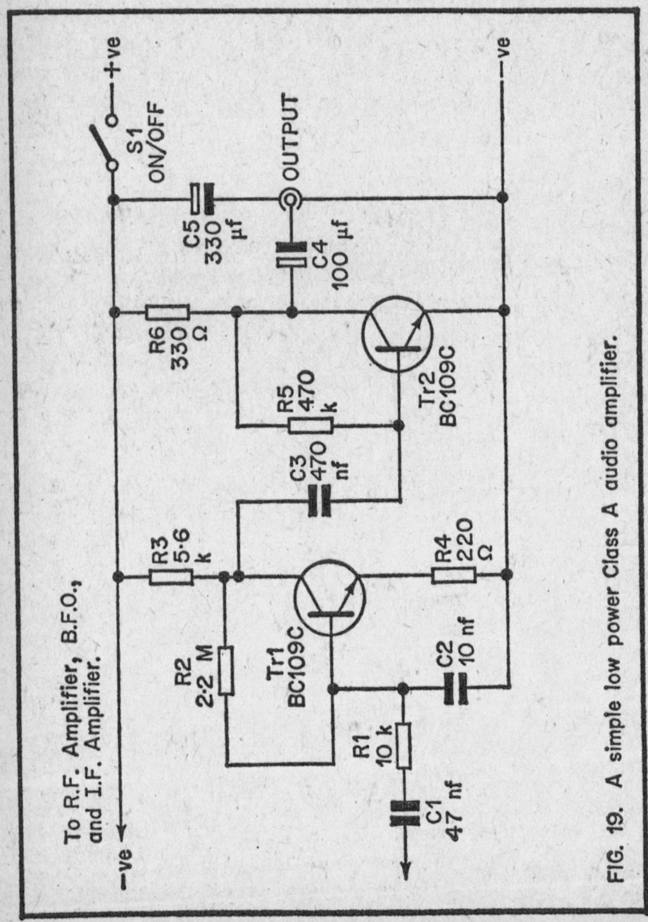

FIG. 19. A simple low power Class A audio amplifier.

Class A Circuit

If the user is primarily interested in headphone reception then the simple discrete low power Class A circuit of Figure 19 should be satisfactory. This is a two stage transistor circuit with both stages incorporating common emitter amplifiers.

Tr1 is the preamplifier transistor, and this has R3 as its collector load resistor and R2 as its base bias resistor. The unbypassed emitter resistor, R4, introduces a certain amount of negative feedback to the preamplifier, and this is necessary in order to reduce the otherwise excessive gain of the circuit. R1 and C2 form an R.F. filter, and aid the stability of the circuit. C1 is the input D.C. blocking capacitor.

The output stage is basically similar to the preamplifier, but no emitter resistor is used, and it is operating at a higher power level. C4 is the output D.C. blocking capacitor, and C5 is a supply decoupling component. S1 is the main on/off switch.

This circuit can be used to feed any normal type of earpiece or headphones satisfactorily, and if loudspeaker reception is required it can also be used to feed a high impedance L.S. This should preferably have an impedance of 40 ohms or more, although satisfactory results may be obtained using some lower impedance speaker units. The maximum output power is rather limited, and so the higher the efficiency of the speaker the better.

B.F.O.

The circuit diagram of a simple B.F.O. having a stabilised supply rail is shown in Figure 20. The use of a stabilised supply conveys two main advantages. Firstly it eliminates the long term drift that can occur as the battery voltage falls due to ageing. Secondly, it prevents modulation of the B.F.O. by changes in the supply voltage due to loading of the power

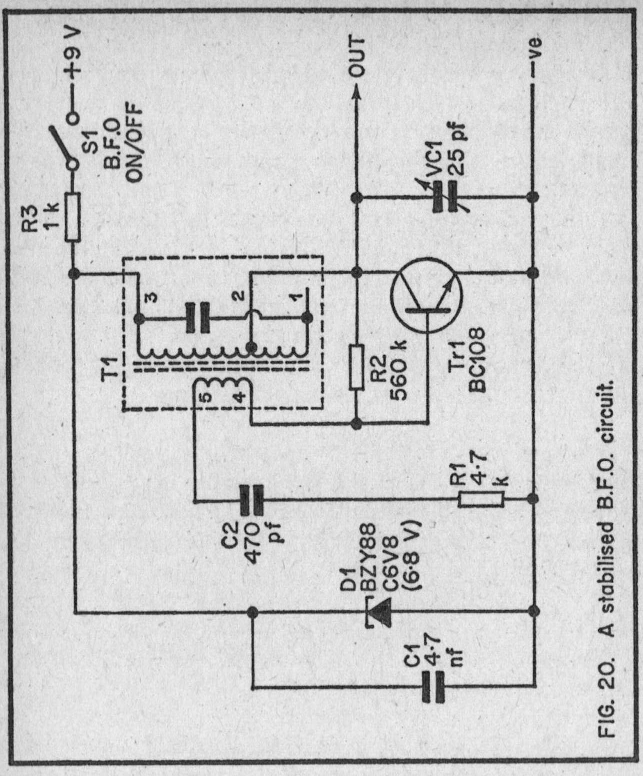

FIG. 20. A stabilised B.F.O. circuit.

supply by the audio amplifier. This is likely to be most noticeable when a Class B audio amplifier is used (such as the two I.C. designs described in this book), and when the battery is nearing exhaustion. This modulation results in a chirping effect on C.W. signals, and a similar effect is produced on S.S.B. signals. In the case of the former this is not too serious, but in the case of the latter it can make it difficult to copy even quite strong and interference free stations. Increased supply decoupling is usually of little value in such cases, and use of a stabilised supply is likely to be far more effective.

In the circuit of Figure 20 the supply rail is stabilised by means of a conventional zener shunt regulator which uses R3 and D1. C1 is an R.F. supply decoupling component. S1 is the B.F.O. on/off switch.

The oscillator circuit is a straightforward inductive feedback oscillator with Tr1 being used in the common emitter mode. R1 is included to reduce the level of positive feedback which is otherwise excessive, and results strong harmonics being produced, as well as numerous other spurious signals. VC1 is a B.F.O. tuning control, and a Jackson type C804 variable capacitor is suitable for use here. The output from the B.F.O. is taken from Tr1 collector, and it should not be necessary to make any direct connection to the I.F. amplifier. A loose coupling of the type which was used in the receiver described in Chapter 1 is all that is really required.

T1 is a Denco I.F. transformer type I.F.T.14/470kHZ, and it is essential that this component is connected with the correct phasing as otherwise the circuit will fail to oscillate. A small information sheet is supplied with each Denco I.F.T., and the numbers next to the leadouts of T1 in Figure 20 correspond to those given on the base diagram of the data sheet which is provided with the I.F.T.14. It is therefore possible to ensure correct connection of this component by reference to Figure 20 and the I.F.T.14 information sheet.

F.E.T./Varicap. B.F.O.

An alternative B.F.O. circuit is shown in Figure 21, and this rather unusual circuit uses a Jugfet in a common source inductive feedback circuit and has variable capacitance (varicap.) diode tuning.

Again a stabilised supply rail is used, and this is really essential here since the B.F.O. is tuned by a voltage which is derived from the supply rail, and therefore, any changes in the supply voltage result in a comparatively large variation in the B.F.O. frequency. The supply is stabilised by a zener shunt regulator which is identical to the one employed in the circuit of Figure 20.

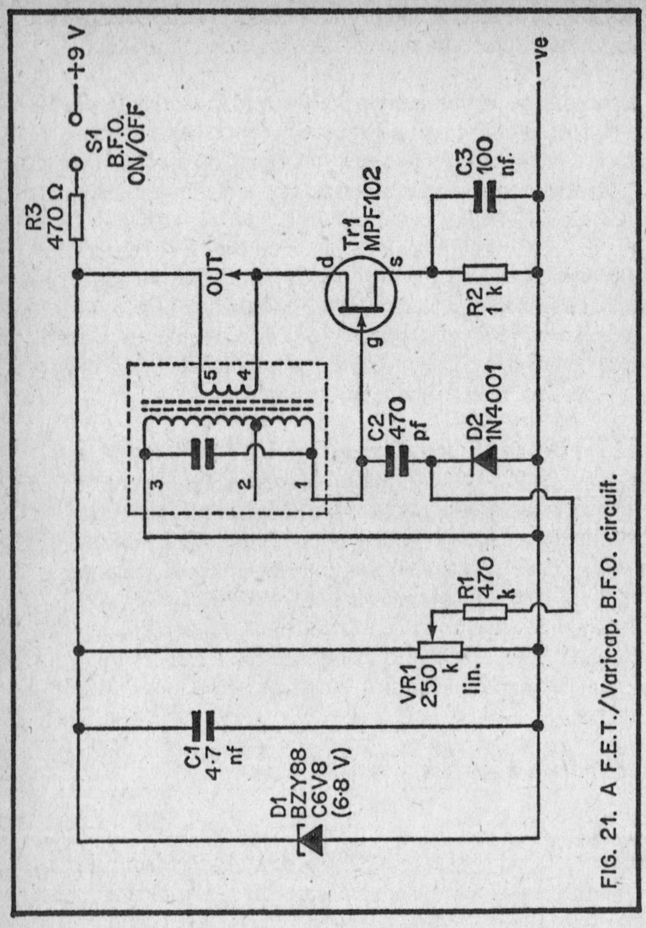

FIG. 21. A F.E.T./Varicap. B.F.O. circuit.

D2 is the tuning diode, and this is coupled to the tuned winding of T1 via C2. VR1 is the B.F.O. tuning control, and a voltage which is variable from zero to the zener voltage is produced at the slider of this potentiometer. This tuning voltage is fed to the varicap. diode via R1. Since the tuning voltage reverse biases D2, no significant current flows in R1, and it can therefore have a fairly high value. This is important, as a low value here would result in heavy loading of the tuned circuit, and this could prevent the circuit from oscillating.

An advantage of this circuit over the one shown in Figure 20 is that it has a lower harmonic output, and it therefore generates fewer spurious signals which will be picked up by the receiver's front end circuitry. It is also a little more stable than the previous B.F.O. circuit.

Although this circuit has a comparatively weak output, again only a loose coupling to the I.F. amplifier will be required. As in the previous circuit, it is essential that T1 is connected with the correct phasing, and T1 is again a Denco I.F.T./ 14/470kHZ component.

As many readers will have realised, the 1N4001 which is specified in the D2 position is not a varicap. diode, but is in fact a 1 Amp. power rectifier. This may seem to be rather unusual, but this choice of component is actually quite practical. Virtually any silicon diode or rectifier will work as a varicap. diode since these all develop a depletion layer when they are reversed biased. This depletion layer is the insulating layer which forms between the two silicon chips at the p – n junction, and it is this layer which prevents the diode from conducting when it is reversed biased. The thickness of the layer varies with the applied voltage, and it increases in depth with increased applied voltage.

When a diode is used as a variable capacitance device it is the depletion layer which forms the dielectric of the capacitor, and the two silicon chips which form the plates. Thus, increased reverse bias voltage results in a thickening of the dielectric, and therefore in a decrease in the devices capacitance.

Proper varicap. diodes are manufactured to very close tolerances so that a certain capacitance, within certain specified limits, will be produced from a given reverse voltage. They are also designed to operate at high frequencies. In this application the characteristics of the varicap. diode are not critical, and the operating frequency is comparatively low. The main requirement is for a fairly high capacitance swing from a rather limited tuning voltage range, and this is provided by the physically large junction of the 1N4001 rectifier at low cost.

Adjustment

The setting up procedure for this receiver is much the same regardless of which building blocks are used, and it is also very similar to the setting up of the receiver described in Chapter 1. It will therefore only be covered fairly briefly here. If the constructor is uncertain about some point of detail, reference to the appropriate part of the previous chapter should clear up any problems.

With an aerial and either phones or a speaker connected to the appropriate sockets it should be possible to receive a few stations with the R.F., mixer, and oscillator coils given approximately the correct adjustment. This merely consists of unscrewing their cores so that about 10m.m. of metal thread protrudes from the top of each coil.

When a station has been located and accurately tuned in, the cores of the I.F. transformers can be peaked. A multimeter having a sensitivity of 20k/volt or more and set to a low D.C. volts range can be used as a tuning indicator. This is connected across VR1 of Figure 17 with the negative lead to chassis and the I.F.T. cores are adjusted for minimum meter reading. If the station is very strong, the voltage across the volume control will become negative with respect to chassis, and it will then be necessary to reverse the polarity of the test periods. The I.F.T. cores are then peaked for maximum meter reading.

The core of the R.F. coil is given any setting which enables the R.F. tuning control to peak stations at any setting of the tuning control. If it is well out of alignment it will not be possible to peak signals at one or other ends of the band, but it is quite possible that no adjustment will be required here.

With the mixer trimmer control set at about half maximum capacitance and the set tuned to a station at roughly the centre of the tuning range, the mixer coil (T2) is adjusted to peak the station.

A multimeter can be used as a tuning indicator for these R.F. adjustments in just the same way as it was for the I.F. adjustments. These adjustments must, of course, be repeated for all three sets of coils. The metal thread which protrudes from the top of each coil will accept 6BA nuts. When the correct settings for the coils have been found, a 6BA nut can be screwed down fairly tightly over this threaded portion so that the core is held in position and will not be accidentally moved during band changing. However, be careful not to upset the settings of the cores when adding these nuts.

The B.F.O. is adjusted by tuning accurately to an A.M. station and then setting the B.F.O. tuning control to a central position. Then the B.F.O. is switched in and the core of the B.F.O. transformer is adjusted to produce a whistle from the phones or speaker. This core is then carefully adjusted to produce the lowest possible beat note at the output.

This completes the setting up procedure and the set is then ready for use. It should perhaps be pointed out that if the passive front end is used, T1 should be regarded as the R.F. coil and T2 as the mixer coil. VC1 then becomes the R.F. tuning control, and VC2 the mixer trimmer control.

The receiver is used in much the same way as that of the previous chapter. The main difference is that there is an additional tuning control in the form of the R.F. tuning one. This is always adjusted to peak received signals, and it will need readjustment each time the main tuning is significantly altered. Small adjustments of the tuning control will not necessitate any readjustment of the R.F. tuning control.

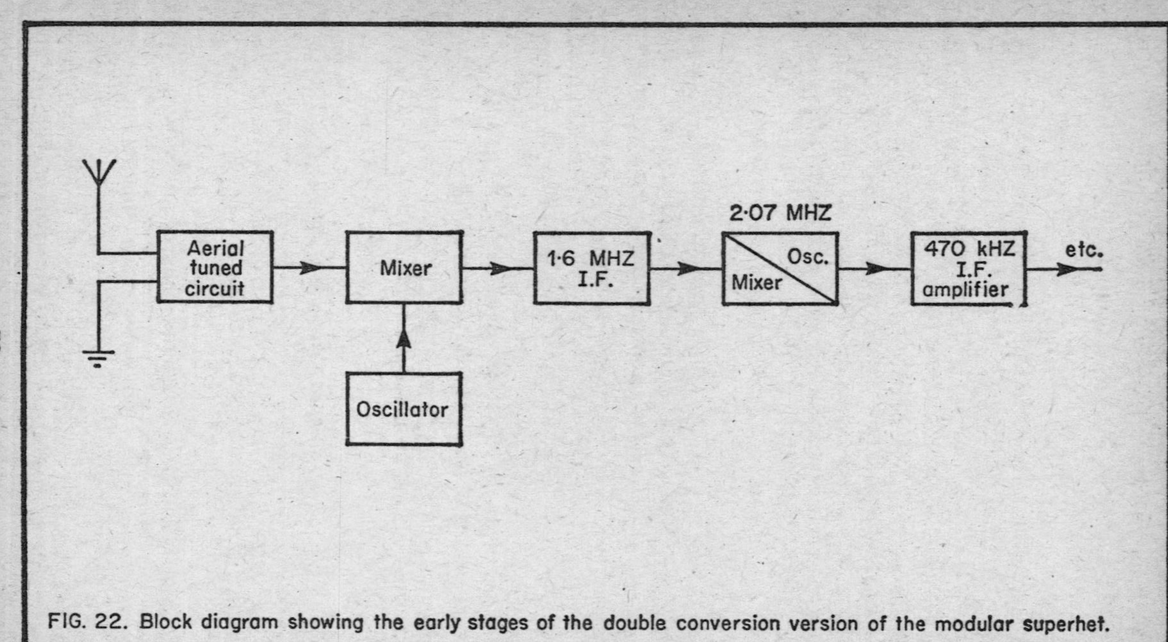

FIG. 22. Block diagram showing the early stages of the double conversion version of the modular superhet.

CHAPTER THREE

DOUBLE CONVERSION RECEIVER

The double conversion approach has already been outlined in the previous chapter, and so the reason for using this type of circuit will not be discussed in detail here. The dual conversion receiver which is described here is based on the circuit modules which were detailed in Chapter 2. The only additional circuitry used in the double conversion receiver is an additional single transistor stage.

The block diagram of Figure 22 shows the early stages of the receiver. At the input of the set any of the front end circuits described in the previous chapter can be used, with no modifications whatever being required. Alternatively a simple passive input circuit using a single tuned circuit can be used, and the relevant circuit appears in Figure 23. This arrangement provides quite good image rejection since this receiver has a first intermediate frequency of 1.6MHZ. This gives an image signal which is some 3.2MHZ away from the main response, and so even a single tuned circuit provides sufficient attenuation of the image response. Of course, the front ends of Chapter 2, with their two tuned circuits, will provide superior image rejection for the additional circuitry and expense that is involved. This will however, be at the cost of decreased cross modulation performance if one of the active front ends is used, or decreased sensitivity if the passive one is employed. The simple circuit . of Figure 23 is thus quite an attractive one from a practical viewpoint.

There is no basic change in the mixer or oscillator circuits, it is only necessary to change the oscillator coil to a colour coded 'White' type and the values of the padder capacitors must be altered. The range 3 padder, which connects to pin 3, requires a value of 340pf, and the range 4 padder, which connects to

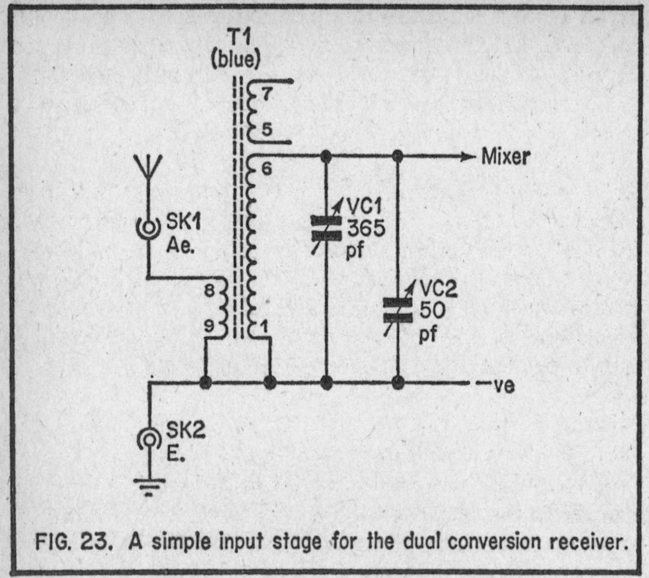

FIG. 23. A simple input stage for the dual conversion receiver.

pin 4, has a value of 960pf. When a 465kHZ I.F. is used there is no need for a padder capacitor on Range 5 and so the relevant coil pin (pin 6) is connected straight to earth. With the increased I.F. a padder becomes necessary and the required value is 2,000pf.

The 1.6 MHZ I.F. merely consists of a 1.6MHZ I.F. transformer and no amplification is provided here. The reason for this is quite simple, and is merely that the I.F. gain of the receiver is already high enough, and any additional I.F. gain would be superfluous. It might even cause problems with overloading.

Some readers might be wondering why an I.F. of 1.6MHZ has been chosen, particularly in view of the fact that the use of an I.F. in the region of 5 to 9MHZ was put forward in Chapter 2. An I.F. of 1.6MHZ is used because oscillator coils for this frequency and I.F. transformers for this frequency are readily available, whereas oscillator coils and I.F.T.s for high intermediate frequencies are not. It is possible to adapt coils and

I.F.T.s which are intended for other frequencies to operate at the desired I.F., but this is not always entirely satisfactory, and is hardly justified when one considers the relatively modest increase in image rejection which is produced. In practice an I.F. of 1.6MHZ will provide good image rejection.

The output of the 1.6MHZ I.F. stage is fed to the second mixer and oscillator circuits. This oscillator has a fixed frequency of 2.07MHZ and it combines with the 1.6MHZ I.F. to produce a second I.F. of 470kHZ (2.07 - 1.6 = 0.47MHZ, or 470kHZ in other words). The output of the second mixer is then fed to a 470kHZ I.F. amplifier, detector, A.G.C. circuit, etc. These circuits are those described in the previous chapter.

Basically then, this receiver is the same as the one of Chapter 2, but the oscillator is slightly modified to produce an I.F. of 1.6MHZ. This I.F. signal is selected by a passive I.F. circuit and then passed on to a second mixer/oscillator stage where it is converted to a conventional 465kHZ I.F. The subsequent stages of the receiver are exactly as before. Thus the high first I.F. gives good image rejection while the low second I.F. gives good selectivity. The sensitivity of the set is much the same as before, and it is actually increased slightly as the gain of the second mixer exceeds the losses through the 1.6MHZ I.F. stage.

Alternative System

It should perhaps be mentioned in passing that there is an alternative type of direct conversion receiver where the first oscillator operates at a fixed frequency and the second oscillator is the tuneable one. This system is outlined in the block diagram which is shown in Figure 24.

Usually with sets of this type the second oscillator has only a restricted tuning range so that each tuning range is rather limited, being perhaps only a few hundred kHZ wide. A few sets of this type have general coverage, but it is more usual to have several tuning ranges with each one centred on an amateur or broadcast band, and there are consequently large gaps in the coverage of such a receiver.

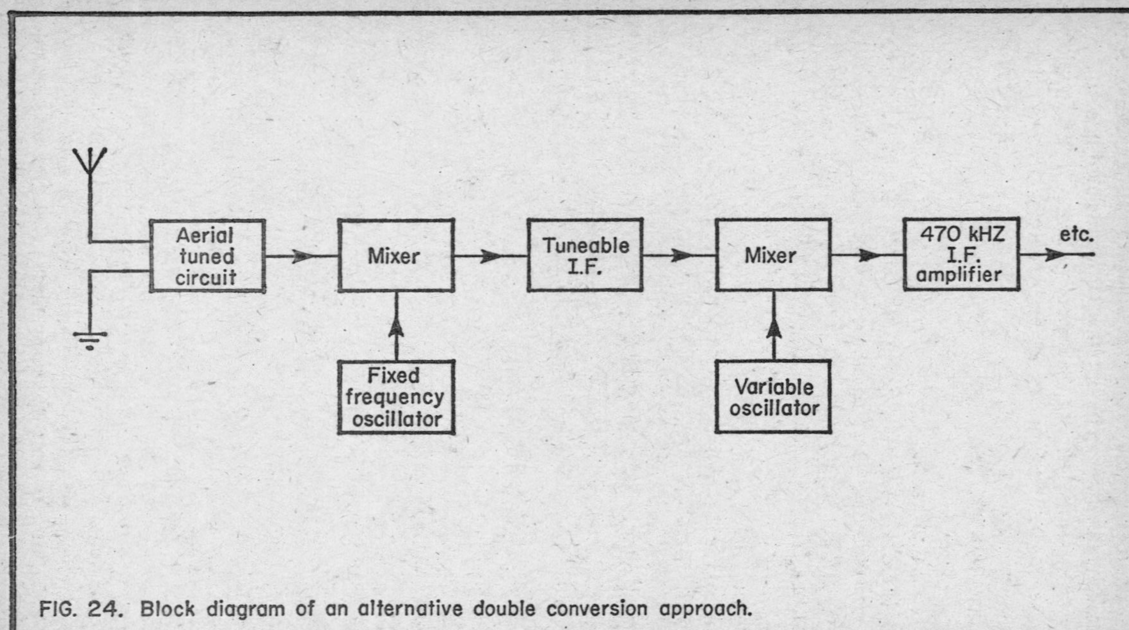

FIG. 24. Block diagram of an alternative double conversion approach.

Sets of this type do have advantages, the primary one being that the first oscillator can be crystal controlled, and therefore excellent stability can be obtained. It is possible to use a crystal controlled second oscillator in a conventional double conversion set, but since the second oscillator is usually at a lower frequency than the first one this conveys comparatively little advantage.

A set of this type will not be described here as it is beyond the scope of most home constructors. One problem is that for every tuning range that is covered a different crystal for the first oscillator is required. Even if suitable crystals could be obtained, they would probably be considered prohibitively expensive by most constructors. The tuning of the receiver is also rather complex since not only is it necessary to have a tuneable aerial tuned circuit and second oscillator, but the first I.F. must also be tuneable.

2nd Mixer/Osc. Circuit

The circuit diagram for the second mixer/oscillator and first I.F. stage appears in Figure 25. This circuit is of the same type as the mixer/oscillator circuit of the receiver which was described in Chapter 1. It may seem a little unusual to use a dual gate MOSFET first mixer and a bipolar second mixer when it has been stated earlier that the latter has a cross modulation performance which is much inferior to that of the former. In fact it is not, as cross modulation is far less of a problem with the second mixer when compared to the first mixer. The reason for this is that the I.F. transformer at the input to the second mixer has a comparatively narrow bandwidth. Therefore only a fairly limited number of signals are presented to the second mixer, and cross modulation is that much less a problem as a consequence of this.

T1 of Figure 24 is a Denco Red Oscillator Coil Range 2T, and this is intended for use as a M.W. oscillator coil. It is suitable for this application because 1.6MHZ lies at the high frequency end of the M.W. band, and so by adding a low value of tuning

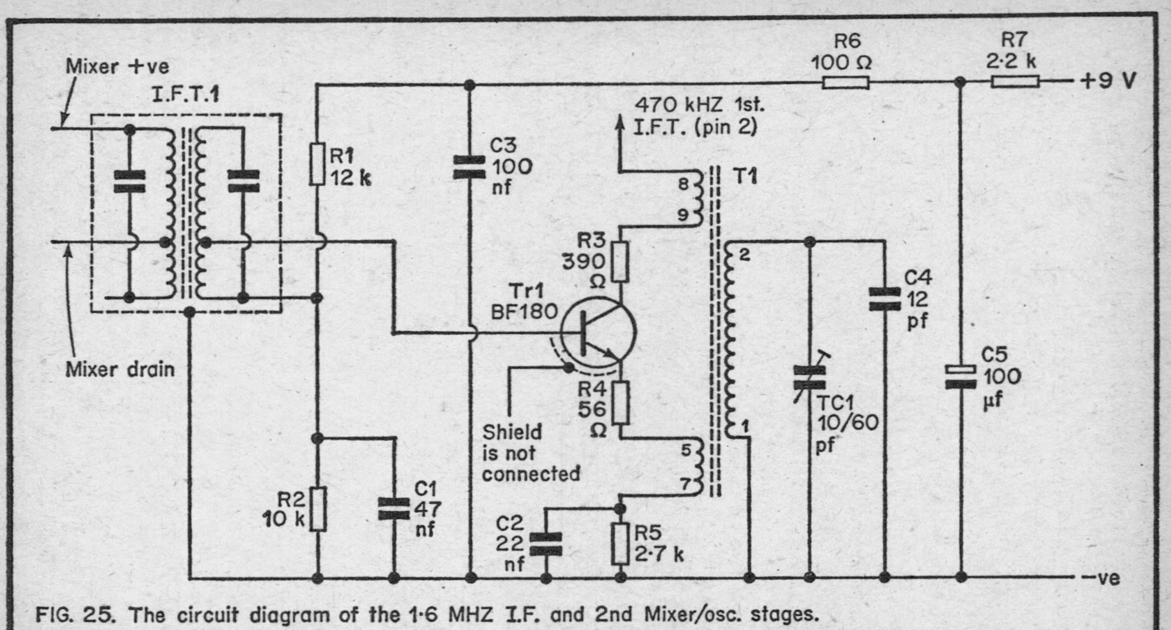

FIG. 25. The circuit diagram of the 1·6 MHZ I.F. and 2nd Mixer/osc. stages.

capacitance across the tuned winding of this coil it can be tuned to produce a 465kHZ output from a 1.6MHZ input.

The supply voltage for the bias circuit for Tr1 is obtained from the main supply line via the decoupling network consisting of R6, R7, C3, and C5. The main supply for Tr1 is obtained from the I.F. amplifier circuit via the primary winding of the first 465kHZ I.F. transformer.

Adjustments

Adjustment of this receiver is very much the same as the procedure used for the sets described in previous chapters. One very important difference is that it will probably not be possible to receive any signals at all until the second oscillator has been set to the correct frequency, or approximately the correct frequency. To do this the core of the second oscillator coil should be adjusted so that approximately 10m.m. of metal screw thread sticks out above the plastic coil former. Then it is a matter of trial and error with TC1 of Figure 24 being tried at various settings, and then the tuning control being adjusted in search of stations. Eventually a setting should be found where the set has a reasonable level of sensitivity.

When this stage has been reached the tuning control should be adjusted in search of a fairly strong signal which seems to have consistant signal strength. Alternatively the output from some form of signal generator (such as the calibration oscillator described in the next chapter) can be used. The latter is preferable as it provides a more reliable signal.

It is then a matter of trial and error again with TC1 being given various settings and then the tuning control being used to accurately tune in the test transmission or signal. The final setting for TC1 is the one which provides the strongest signal. TC1 cannot be peaked in the same way as an R.F. or I.F. tuned circuit because it affects the reception frequency, in much the same way as the first oscillator does.

When this adjustment has been satisfactorily completed the receiver is then adjusted in the normal manner.

Finally, when all other adjustments have been completed, the 1.6MHZ I.F. transformer is peaked. Make sure the set is very accurately tuned to the test signal before adjusting this component. It has two cores, and the form of construction is the same as that used for the 465kHZ double tuned I.F. transformers.

Construction of the unit should present no real difficulties, but it is advisable to completely screen this part of the set from the other circuitry. The reason for doing this is simply that there is a certain amount of harmonic output from the second oscillator and these can be picked up by the aerial circuitry unless suitable screening is used. Even the fundamental frequency of the second oscillator (2.07MHZ) lies within the frequency coverage of the set (Range 3T).

One way of screening off the second mixer/osc. circuitry is to build it into a metal box within the main casing of the receiver. Diecast aluminium boxes are ideal for applications such as this where good screening properties are required. It is necessary to drill small holes in the box at strategic points so that it is possible to adjust the cores of the I.F.T. and also TC1. The positive supply rail can be taken into the box via a 1nf feed through ceramic capacitor.

Bandswitching

It has been assumed so far that whether the constructor is going to build the single or the double conversion receiver, plug in bandchanging will be used. However, many constructors would no doubt prefer to have the convenience of bandswitching even though this is more difficult to achieve than one might think.

Fortunately it is not necessary to switch both sides of each coil winding, and it is only necessary to switch the non-earthy end, or 'hot' end as it is often called. The earthy or 'cold' ends are merely connected in parallel. Note that the cold end of a winding very often does not connect direct to earth, but is instead connected to earth via an earth return capacitor. Occasionally neither end of a winding will be found to be earthed (such as the collector winding of T2 in Figure 3). Even in cases such as this it is only necessary to switch one side of the wiring.

In deciding which end of a winding should be switched just follow this simple rule. If one end of a winding is earthed then put the switching in the other side, if one end of a winding is coupled to a terminal of a transistor, it is this side of the winding which is switched. Thus the wavechanging for the circuit of Figure 3 would be as shown in the skeleton circuit of Figure 26.

It is possible to slightly simplify the bandchanging by connecting the aerial coupling windings in series and not bothering with any switching at all in this part of the circuit. This will, however, result in a slight loss of efficiency.

It is essential that a well planned layout is used as otherwise loss of performance, instability, and other problems are likely to result. Neat appearance and convenience are of secondary importance, and short wiring is the main requirement.

Rather than mount the coils in B9A holders it is probably better to chassis mount them. Each coil has a OBA plastic thread at the top and is supplied with a plastic OBA nut. The coils are thus easily mounted on the chassis or a mounting plate with only a single OBA mounting hole (¼in. or 6.5m.m. diameter drill) being required for each coil. The plastic thread is easily stripped from the former and so the mounting nut be taken beyond finger tightness. Another slight problem is that there is a tendency for the polystyrene coil former to melt if soldered connections are made direct to the pins of the coil. Soldered connections to these pins must therefore be completed quickly so as to avoid damaging the coils. On the other hand, make quite sure that no dry joints are produced and do not sacrifice the quality of the joint for the sake of speed.

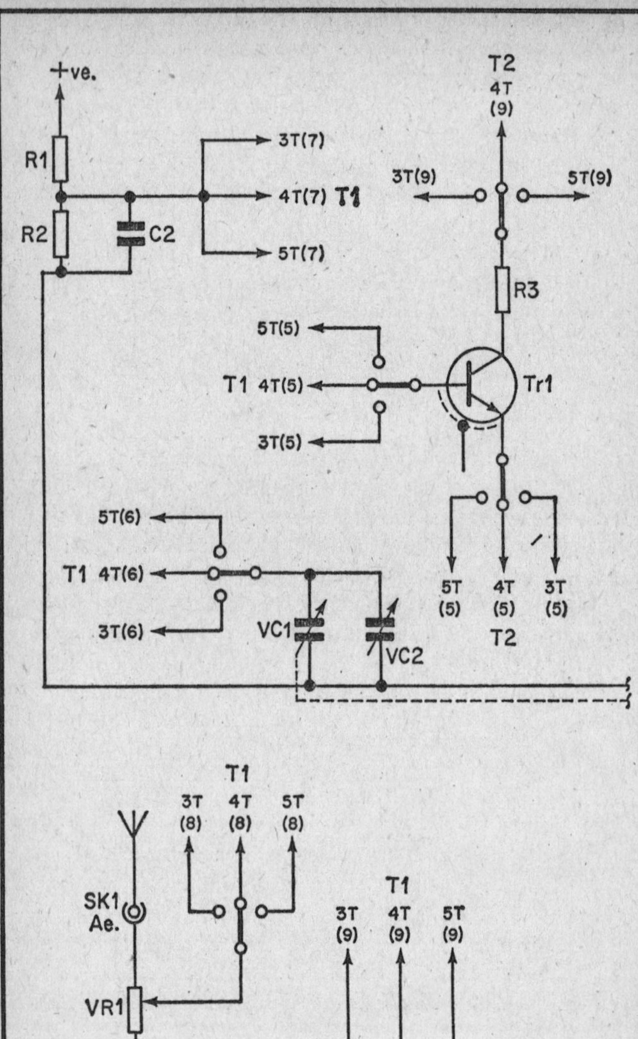

FIG. 26. Circuit details for bandswitching the circuit of Fig. 3.

3T(3)
T2 4T(3)
5T(3)
To I.F.T.1

T2
3T (1)
4T (1)
5T (1)
T2
3T (3)
4T (4)
5T (6)
C4
C5
VC3
C6
T1
3T (1)
4T (1)
5T (1)
T2
3T (7)
4T (7)
5T (7)
C3
R4
-ve.

OR

SK1
Ae.
VR1
T1
3T
4T
5T
8
9
-ve.

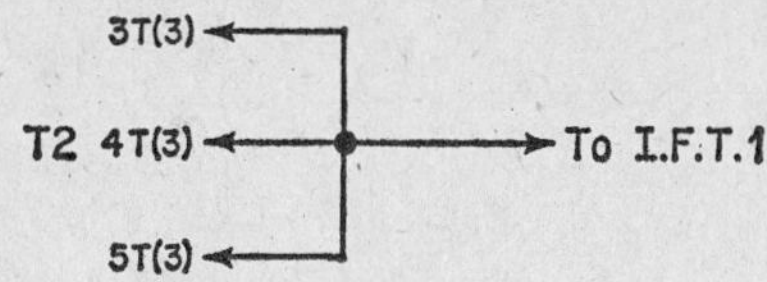

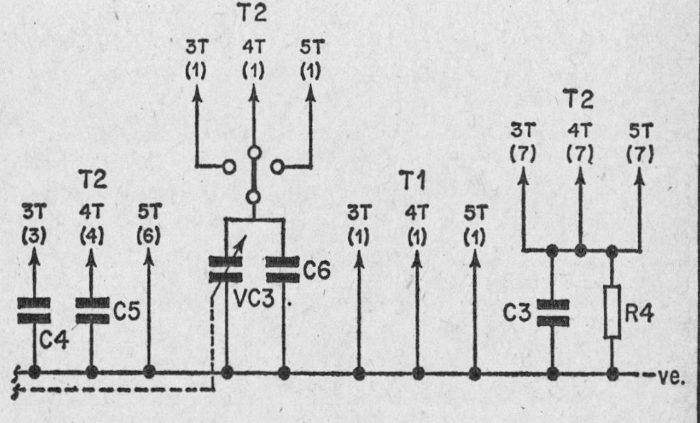

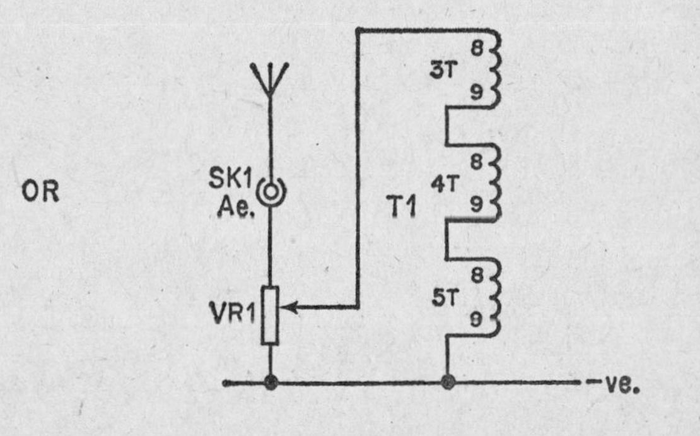

Note that shortness of wiring is more important on the higher frequency ranges than it is on the lower frequency ones. It is inevitable that the wiring to some coils will be longer than the wiring to others, and so the layout should be arranged in such a fashion that the shortest wiring connects to the Range 5T coil.

Apart from the fact that long wires are likely to cause a greater loss of efficiency on Range 5T there is the additional problem that long leads in the wiring associated with the tuned windings will significantly increase the inductance of these windings. This can result in a loss of the extreme H.F. coverage of the receiver and it can also cause problems with alignment.

If bandswitching is used it is possible to eliminate the mixer trimmer control by adding a 10/40pf (approx.) trimmer capacitor across the tuned winding of each mixer coil. It is also necessary to remove the fixed capacitor connected across the oscillator tuning capacitor, and then add a 10/40pf trimmer across the tuned winding of each oscillator coil.

With the aid of a signal generator the oscillator trimmer is adjusted to set the correct upper frequency limit of one tuning range. Then the core of that oscillator coil is adjusted to produce the correct lower frequency limit on that tuning range. Then the trimmer is readjusted to produce the proper limit of H.F. coverage once again, and this whole procedure is repeated several times until the correct coverage (or very nearly so) is obtained.

Then the set is tuned to the high frequency end of the range and the mixer trimmer is adjusted to peak sensitivity. Next the set is tuned to the low frequency end of the range and the core of the mixer coil is adjusted to peak sensitivity. The receiver is then returned to the high frequency end of the tuning range once again and the mixer trimmer is repeaked. This procedure is repeated several times until no further improvement in overall sensitivity can be obtained.

All three ranges must be adjusted in this way.

This system does not provide such good results as the use of a mixer trimmer control as the latter enables the mixer tuned circuit to always be perfectly peaked at any setting of the tuning control. However, any loss of performance will only be very small, and this system is more convenient since one of the tuning controls is eliminated.

It is recommended that a separate wavechange switch is used for the R.F. amplifier circuitry in the same way that a separate tuning control is used, and for the same reasons. It is possible to use a single switch for bandchanging, but careful and extensive screening would be necessary.

CHAPTER FOUR

ADD-ON CIRCUITS

This final chapter will deal with a number of add-on circuits which can be used to aid and improve reception using the receivers described in Chapters 2 and 3. Many of these circuits can be used with other receivers and are of general interest. A surprisingly large number of add-on circuits are provided, and no excuse is made for this fact. The receivers described in the two previous chapters are capable of extremely good results and have high levels of performance, but results can, nevertheless, be greatly improved by some additional circuitry. Not all the circuits described here actually improve performance at all, and an example of such a circuit is the one for an 'S' Meter. Units such as this can be very useful though, and most S.W.L.s and amateurs would probably consider them a virtual necessity.

Q Multiplier

One area in which the receivers have a level of performance which leaves a lot to be desired is the degree of selectivity which they provide. Ideally a bandwidth of about 5 to 6kHZ is needed for A.M. reception, with a response which falls off very rapidly outside these limits. In order to achieve this level of selectivity using ordinary I.F. transformers it is necessary to use an I.F. of even less than 465kHZ. Some designs use a triple conversion circuit with a final I.F. of only 100 or 85kHZ in order to obtain high selectivity, and excellent results on A.M. can be obtained in this way.

However, for S.S.B. reception a bandwidth of only about 2.5kHZ is required, and for C.W. reception the ideal bandwidth is even less. Modern receiver designs tend to use ordinary I.F.s with some form of I.F. filter rather than bother with the complication of an extra converter stage.

The most sophisticated of modern receivers uses a different crystal filter for each mode of reception, and each filter has a response characteristic which is moulded to suit the appropriate mode. This method is beyond the scope of most amateur constructors because of the high cost and complexity which is involved.

Variable selectivity can be obtained by using quite a simple device known as a Q multiplier. This will not give a level of performance which equals a set of high quality crystal filters, but it will provide excellent results for very little outlay. The circuit diagram of a Q multiplier is shown in Figure 27.

A Q multiplier relies on the fact that an ideal tuned circuit has a very narrow bandwidth and a steep sided response. Practical tuned circuits fall well short of perfection though due to the

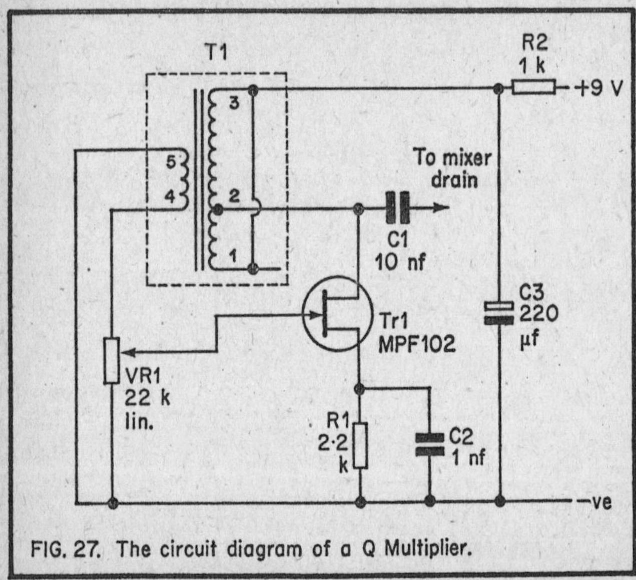

FIG. 27. The circuit diagram of a Q Multiplier.

resistance of the wire from which the coil is constructed. It is this resistance which introduces losses into the coil and results in a widening of its bandwidth and a much more gradual fall off in the passband. The effectiveness of a coil is known as its Q, and the purpose of a Q multiplier is to boost the Q of an ordinary tuned circuit. By connecting this tuned circuit in the I.F. amplifier circuitry of a receiver a considerable improvement in selectivity can be obtained.

The way in which a Q multiplier increases the Q of a coil is really very straightforward. The tuned circuit is connected at the output of an amplifier circuit, and some of the energy in the coil is coupled to the input of the amplifier. There must be zero phase shift between the input to and output from the tuned circuit. The effect of the circuit then is that it tends to compensate for the losses in the coil. The amount of feedback which is applied over the circuit is made variable so that the effective Q of the coil (and therefore the bandwidth of the receiver) can be altered. With little or no feedback used the selectivity of the receiver will be little affected by the Q multiplier. On the other hand, with the feedback adjusted so that all the losses in the coil are compensated for, a very high Q and thus also level of selectivity will be attained. Intermediate feedback levels will provide intermediate degrees of selectivity. Therefore it is possible to adjust the selectivity to suit the reception mode.

Note that if the level of feedback is overdone even slightly, the Q multiplier will break into oscillation. Its effect will then be very much like switching in the B.F.O. and it will not provide the desired effect.

In the circuit of Figure 27 Tr1 is used as a common source amplifier and it has the tuned circuit as its drain load. Positive feedback is applied to the gate of Tr1 via the secondary winding of T1 and VR1. The latter controls the amount of feedback and acts as the bandwidth control. T1 is a Denco I.F.T.14/470kHZ component.

Adjustment

When initially testing the circuit, accurately tune to a strong A.M. station before significantly advancing the level of feedback using VR1. When VR1 is well advanced the Q multiplier should begin to oscillate, and this will produce a beat note from the phones or speaker. With VR1 advanced just beyond the threshold of oscillation, adjust the core of T1 for zero beat. The unit is then in alignment with the I.F. amplifier, and it is ready for use.

It should be found that a vast improvement in selectivity is obtained, and with the feedback level carefully set just below the threshold of oscillation the bandwidth should be so narrow that it is only suitable for proper reception of C.W.

Crystal Filter

Most crystal filter circuits are rather expensive to construct and difficult to align accurately, but the simple circuit which is shown in Figure 28 is quite easy to set up, and as only a single crystal is used it is comparatively inexpensive.

The bandwidth obtained with this arrangement is just about adequate for A.M. reception, although it is really a little narrower than would ideally be used. For S.S.B. reception the bandwidth is perhaps a little wider than the ideal, but the bandwidth is a good compromise for these two reception modes. The passband is considerably wider than is really necessary for C.W. reception, but it is possible to overcome this by using an audio C.W. filter. Such a circuit will be described later.

The crystal is the circuit element which provides the very sharp selectivity, and this is a piezo electric device which has two resonant frequencies. At the series resonant frequency the crystal has a very low impedance, and at the parallel resonant frequency it has a very high impedance. These two frequencies

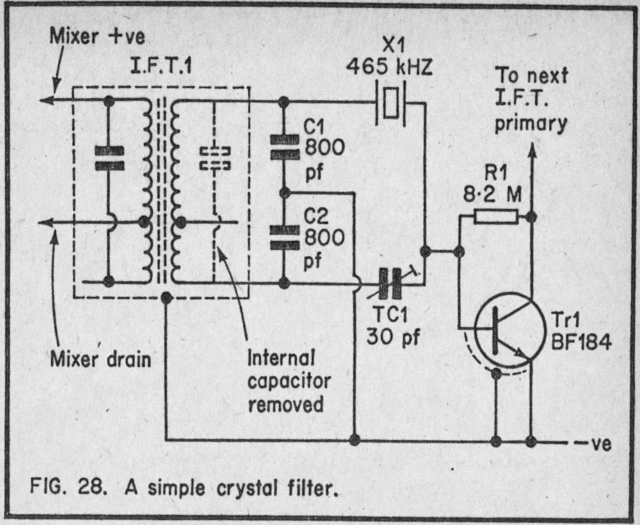

FIG. 28. A simple crystal filter.

are very close together, usually being something like a few hundred HZ to a couple of kHZ apart. The series resonant frequency is the lower of the two.

This sudden transition from a low to a high impedance with only a very small increase in input frequency means that a crystal has an extremely high Q value. The Q of a crystal is usually something in the order of several thousand, whereas the Q of a tuned circuit in a conventional I.F.T. is normally something in the region of one hundred. This very high Q enables even a simple crystal filter to achieve a very narrow bandwidth having an extremely rapid roll-off outside the required passband. This filter therefore provides really excellent results.

Basically the crystal is used in a form of bridge circuit, and the output of the bridge is fed to a simple common emitter amplifier. The purpose of the amplifier is to compensate for the losses through the filter proper. R1 is given a fairly high value so that this stage has only a fairly low level of gain and a low noise level. The level of gain can be increased, if required, by decreasing the value of R1 somewhat.

I.F.T.1 is a Denco I.F.T.18/465KHZ type, and this needs to be modified slightly. This modification is necessary because a centre tapped secondary winding is needed, and to the best of the authors knowledge no suitable ready made component is available. The tapping on the secondary of the Denco I.F.T.18/465KHZ component is unsuitable as it is not a centre tap.

This problem is overcome by removing the internal capacitor connected across the secondary winding, and substituting two external series capacitors of twice the value. These external capacitors are C1 and C2 of Figure 28. The combined capacitance of these two components is equal to that of the removed internal capacitor, and so the resonant frequency of the secondary winding is not significantly altered. The junction of C1 and C2 forms a capacitive centre tap, and this is just as effective as using an inductive centre tap.

The removed tuning capacitor has a value of 400pf, and so each of the new capacitors requires a value of 800pf. It is unlikely that components of this value will be available, and so it will be necessary to use two capacitors wired in parallel to make up the required value. The author used a 470pf and a 330pf capacitor to make up this value. Use good quality close tolerance (preferably 2% or better) components.

It is an easy matter to remove the screening can from the I.F.T. so that access to the capacitor can be obtained. It is merely necessary to push outwards two small indentations at the base of the can. The screening can is then slipped off and the capacitor is snipped out using a pair of wire clippers. The capacitor is mounted on the inside of the base of the I.F.T., and care must be taken here as otherwise the leads to the secondary winding could easily be cut (but not so easily restored). Also make sure that the right capacitor is removed as there is an identical one connected across the primary winding. When this task has been completed the screening can is replaced.

Adjustment

Start with TC1 set at about half capacitance and then try to tune a few stations on the receiver. It will probably be found that it is possible to tune each station twice, with the two tuning points being very closely spaced. This effect will be more noticeable if the unit is connected to an R.F. signal generator rather than an aerial, as there will then be only a single signal to tune to. One of these responses will be relatively broad, and will be affected by adjustment of the I.F.T. cores. The other will be much more sharp and will be unaffected by adjustments made to the I.F.T.s.

The cores of the I.F.T.s must be adjusted to move the broad response towards the sharp one. When the two virtually coincide, accurately tune in the signal on the sharp response and then peak the I.F.T.s in the normal way.

It may happen that the two responses virutally coincide without any adjustment being required. The set will then be found to have a rather unusual passband with two very closely spaced peaks. The response will have a sharp fall off on one side and a much more gradual attentuation curve at the other side. Tune the receiver to peak a signal in the response peak which is on the same side of the response as the rapid roll-off. The I.F.T. cores are then adjusted to peak this signal.

It should perhaps be explained that the broad response is the one produced by the I.F.T.s, and the narrow one is the response produced by the crystal. Obviously, it is necessary to have the two responses coinciding for satisfactory results to be obtained.

Adjustment of TC1 should be found to alter the shape of the response quite noticeably, although with some crystals it may be found to have little effect. This trimmer is given any setting which produces a reasonably symmetrical response.

If a filter in/out switch is required, this can be achieved by connecting a S.P.S.T. switch across the crystal. When this switch is closed it short circuits the crystal and the response is then that produced by the I.F.T.s alone. When the switch is open the filter operates normally.

It is not essential for the crystal to have an operating frequency of precisely 465kHZ, and anything from about 460 to 470kHZ should be satisfactory. The author employed a 468kHZ crystal in the prototype.

Mechanical Filters

There are currently available a number of mechanical and ceramic filters which can be used instead of either one or both of the double tuned I.F. transformers used in the receiver in order to produce increased selectivity. By using the narrow band versions of these filters it is possible to obtain quite a high degree of selectivity. One such filter is the Toko MFH-41T.

'S' Meter

An 'S' meter or signal strength meter is very useful as a source of comparative signal strengths, and it can also be used as a tuning meter. The circuit diagram of a 'S' meter unit for use with the I.F. Amplifier circuit of Figure 17 is shown in Figure 29.

This circuit takes its output from the A.G.C. feedback loop of the I.F. amplifier, and so the input voltage decreases with increases in signal strengths of received transmissions. I.C.1 is used as a simple D.C. amplifier having a voltage gain of about 4 times. The quiescent input voltage from the I.F. amplifier produces an output voltage from I.C.1 of about 6 volts. VR1 is connected as part of a potential divider across the supply lines, and this is adjusted so that the voltage at its slider is equal to the output voltage of the I.C.

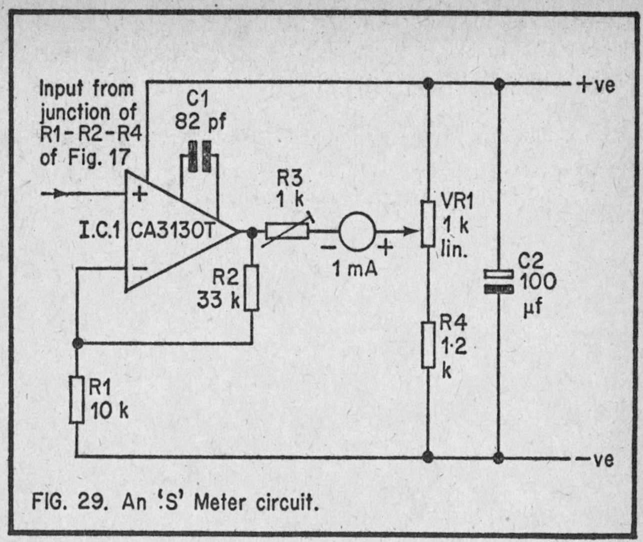

FIG. 29. An 'S' Meter circuit.

The meter is a standard 1mA. type which together with R3 forms a simple voltmeter circuit. When an input signal is present the A.G.C. voltage will reduce and this causes a larger voltage reduction at the output of I.C.1. This causes the voltage at the negative terminal of the meter to be less than that at the positive one, and in consequence a positive deflection of the meter is produced.

The A.G.C. voltage falls with increasing signal strength, and so the level of meter deflection is proportional to the signal strength of received signals.

R1 and R2 are a negative feedback loop which set the voltage gain of I.C.1 at the required level. C1 is the compensation capacitor for I.C.1 and C2 is a supply decoupling capacitor. The input impedance of the CA3130T is in the order of 1.5 million meg. ohms, and so it does not significantly load the A.G.C. circuitry and no modification of the I.F. amplifier is required.

Adjustment

Before switching on the set, adjust R3 to insert maximum resistance into circuit and adjust VR1 so that virtually the full supply rail potential is present at its slider. Disconnect the aerial from the receiver and make sure that the meter is properly mechanically zeroed.

When the receiver is switched on there should be an immediate positive indication on the meter. VR1 is then adjusted to zero the meter. The aerial is then reconnected and the 'S' meter is tested. It will probably be found that even the strongest signals produce only a relatively low meter reading. This is corrected by adjusting R3 for decreased resistance. This is given a final setting which produces virtually f.s.d. on extremely strong signals, but which does not produce readings off the end of the scale on the very strongest signals.

Noise Limiter

Noise spikes, which can be caused by man made and natural sources, can occasionally make it difficult to copy a signal properly. There is no way of completely eliminating this problem, but it is possible to minimise it by using a noise limiter circuit. The circuit diagram of a simple noise limiter is shown in Figure 30, and this circuit is simply added in the audio output of the receiver.

This really just consists of a simple clipping circuit with D1 and D2 being used as low voltage (about 0.5 volt) zener diodes. R1 is the load resistor for these. D1 clips to positive going output half cycles and D2 clips the negative going ones.

In use the audio output level is adjusted so that the desired signal just about reaches the clipping level. Then any loud noise spikes which are received will be clipped at the output to a level which is no more than the peak level of the desired

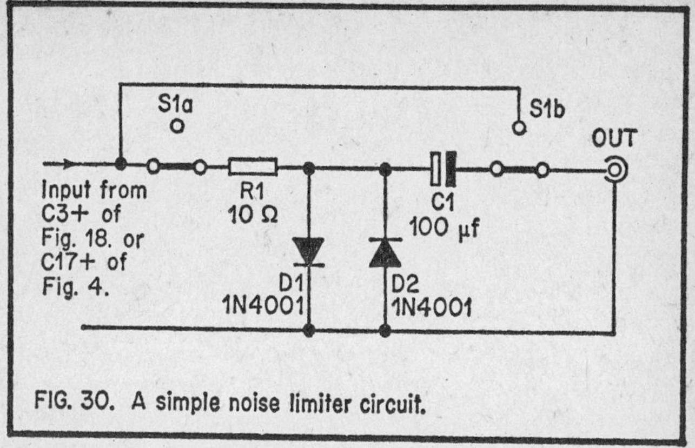

FIG. 30. A simple noise limiter circuit.

signal. This effect is illustrated in Figure 31. Thus, even if the noise spikes have an amplitude of one hundred times that of the required signal, at the output their levels are equal.

Obviously only a limited improvement in intelligibility can be produced in this way, but the improvement is well worth-while for the very small increase in cost and complexity which is imparted by the additional circuitry. It can be particularly useful for headphone listening where even occasional noise spikes can be very troublesome due to their almost deafening volume.

S1 is the noise limiter on/off switch, and in the off position it simply bypasses the limiter circuitry, C1 is the output coupling capacitor.

There is another form of noise reduction circuit known as a noise blanker, and although this is more effective than a noise limiter it is not very popular due to its greatly increased cost and complexity. For this reason a practical circuit will not be given here.

Basically a noise blanker consists of a fairly insensitive receiver with a bandwidth which is extremely wide (usually several MHZ). Ordinary radio signals will not be received very strongly,

7

but noise spikes which are very strong and extend over a wide range of frequencies will be.

The output of the receiver is used to mute the I.F. stages of the main set in the presence of a strong signal, and so the receiver is muted whenever a strong noise spike comes along. Thus the usual crack sound of a noise spike is not merely limited in amplitude, it is completely eliminated.

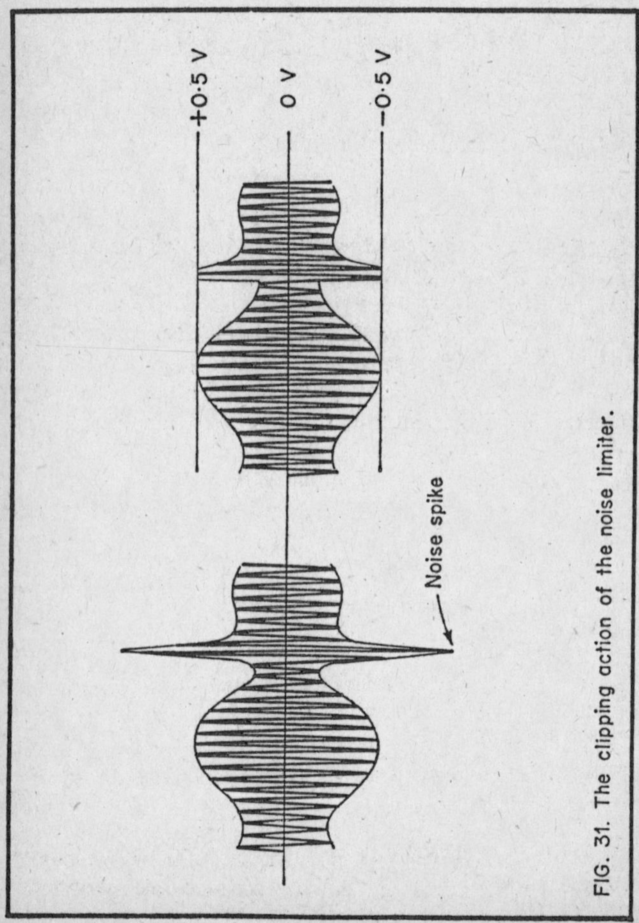

FIG. 31. The clipping action of the noise limiter.

C.W. Filter

Theoretically an infinitely narrow bandwidth is all that is needed for proper reception of C.W. signals since only a single reception frequency is involved. In practice though, even a bandwidth of a few HZ is impractically narrow since tuning would then be so sharp as to be virtually an impossibility, and the stability of the set would need to be higher than can be attained in a practical situation. Therefore a bandwidth of about 100HZ to about 600HZ is usually accepted as being the optium range for C.W. reception.

It is not at all easy to achieve such a narrow bandwidth in the I.F. stages, although it can be done with expensive crystal filters, and a Q multiplier can come close to the optium bandwidth. However, a simple way of obtaining a narrow bandwidth for C.W. reception is to use an audio filter. The circuit diagram of Figure 32 shows such a unit, and this will allow frequencies at or close to 1.5kHZ to pass, and will reject other frequencies.

It should perhaps be explained that this method of using an audio filter is not quite as effective as using an I.F. filter having the same bandwidth. The reason for this is simply that the audio filter has two responses, in effect anyway, since an I.F. signal either 1.5kHZ above or 1.5KHZ below the B.F.O. frequency will produce a 1.5kHZ audio output. However, very good results can be provided by an audio filter, and many highly sophisticated communications receivers incorporate such circuits.

The one shown in Figure 32 is based on a T notch filter, or a twin T filter as it is sometimes alternatively known. The components in this network are C1 to C4, and R1 to R3. A T notch filter gives a narrow rejection notch at (and near) to its operating frequency, and allows other frequencies to pass. This is, of course, just the opposite of what is required here!

Therefore the filter has been connected to a simple common emitter amplifier using Tr1, and it is connected between the

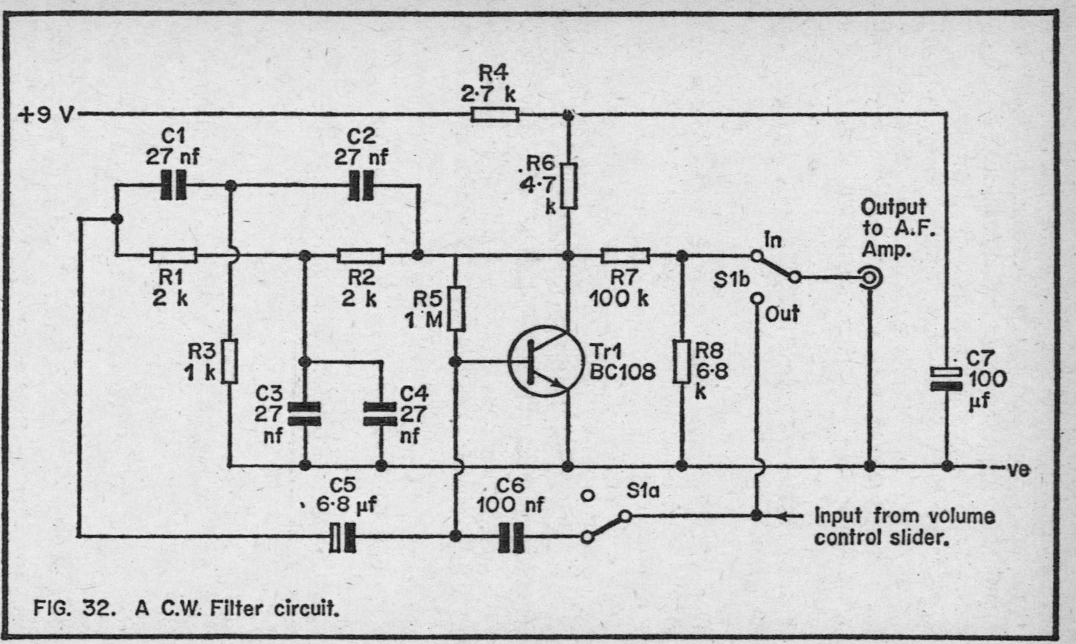

FIG. 32. A C.W. Filter circuit.

collector and base of this transistor. These two points are 180 degrees out of phase, and negative feedback is provided through the T notch filter at frequencies which are permitted to pass. On the other hand, at and close to its operating frequency it will not provide a negative feedback path.

Thus at most frequencies the circuit has only about unity voltage gain due to the large amount of negative feedback, but at the operating frequency of the filter the amplifier has virtually its full gain (about 40dB.) due to the lack of feedback through the T notch filter..

The full additional gain of the filter is not really required, and in fact it would almost certainly result in instability if no steps to reduce the gain were taken. Therefore an attenuator consisting of R7 and R8 has been added at the output of the filter.

S1 is the filter on/off switch, and when it is in the 'off' position it bypasses the filter. R4 and C7 are a supply decoupling network. C5 and C6 are D.C. blocking capacitors.

Variable Notch Filter

Probably the most frequent type of interference that is encountered on the S.W. bands is heterodynes caused by two A.M. stations on very close frequencies, or a heterodyne between an A.M. station and the B.F.O. of the set. It is possible to virtually eliminate this type of interference by using a variable notch filter.

This type of filter allows most frequencies to pass, but blocks signals which lay close to or at its operating frequency. It is made tuneable so that it can be adjusted to block an interfering heterodyne.

The basic T notch filter used in the previous chapter may appear to be an obvious choice for use in this application but unfortunately it is not really suitable. The problem is that it

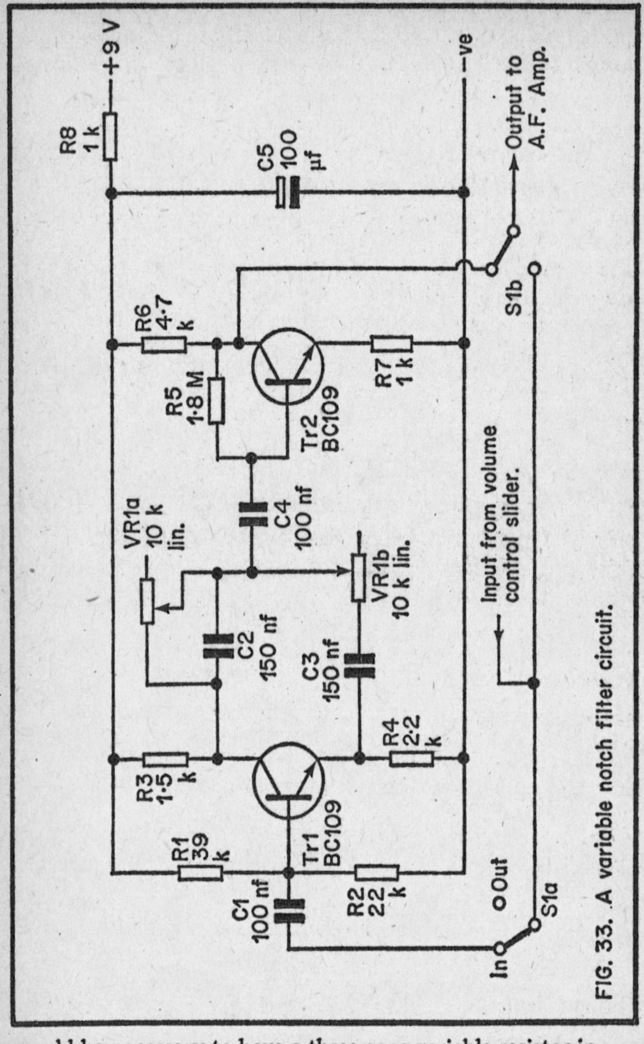

FIG. 33. A variable notch filter circuit.

would be necessary to have a three gang variable resistor in order to make this filter tuneable, and a suitable component does not appear to be available. It is necessary to use an active filter, and the circuit diagram of a tuneable notch filter is shown in Figure 33.

Tr1 is used as a phase splitter with two out of phase signals being produced at the collector and emitter. The collector output is fed to a phase shift network consisting of VR1a and C2 while the emitter output is fed to a phase shift network consisting of C3 and VR1b. At one frequency these signals will both undergo the same degree of phase shift. The outputs of the phase shift networks are mixed at the junction of the two potentiometer sliders, and so an input signal at the frequency where the two shifters produce the same phase change will be cancelled out to a large degree. At other frequencies little or no cancelling takes place. The circuit thus provides the required rejection notch.

The filter is tuned by adjustment of VR1, and this provides a frequency range of approximately 100HZ to 5kHZ.

Even though the filter is an active type, there is a substantial loss through it, and it needs to feed a high impedance in order for optium results to be obtained. Tr2 is used as a buffer amplifier at the output, and this provides a high load impedance for the filter and compensates for the filter losses.

R8 and C5 are a supply decoupling network. C1 provides D.C. procking at the input to the filter. S1 is the filter in/out switch, and when it is in the 'out' position it simply bypasses the filter circuitry.

Note that in order for best results to be obtained it is necessary to use 5% tolerance (or better) components in the C2 and C3 positions. Incidentally, this also applies to the resistors and capacitors in the T notch network of the previous circuit.

Good results should be obtained with the circuit as shown, but if R3 is replaced with a fixed 680 ohm resistor and a 2.2k pre-set wired in series, it will then be possible to adjust the preset for optium rejection.

Using the filter is extremely simple, and it is just a matter of adjusting VR1 to minimise any offending heterodyne.

Product Detector

Although quite good results on S.S.B. can be obtained using a B.F.O. loosely coupled into the I.F. amplifier stages of the receiver, a higher quality output can be obtained by using a device known as a product detector. This type of circuit is also more convenient to use as it improves the sets immunity to overloading on strong S.S.B. signals. Another problem with the loose coupling method of B.F.O. injection is that if the 'S' meter circuit is fitted to the receiver, this circuit will not function properly in the C.W./S.S.B. mode. This is because the B.F.O. signal will cause a strong positive deflection on the meter meter. By using a screened B.F.O. and product detector this can be overcome.

The circuit diagram of a product detector plus a stabilised B.F.O. stage is shown in Figure 34. A product detector is a mixer circuit of the same type that is used in the frequency converter stage of a superheterodyne receiver. However, instead of converting the R.F. signal to the intermediate frequency, it converts the I.F. signal to an audio frequency.

When a B.F.O. signal loosely coupled into the I.F. stages is used for S.S.B. reception the audio quality usually leaves a lot to be desired. This is because not only is the B.F.O. signal reacting with the S.S.B. signal components to produce an audio output, but the sideband components are reacting with one another to produce an audio output. This signal can be clearly heard if the B.F.O. is switched off. This reaction between the sideband components is undesirable because the audio output it produces was not present in the modulation signal at the transmitter, and it is therefore a form of distortion.

By using a product detector this distortion is largely eliminated since the audio output is produced by the B.F.O. reacting with the sideband components, and not by these components reacting with one another. In practice some reaction of this kind does occur, but only on a comparatively limited scale. If the B.F.O. is switched out when a product detector is being used, very little output will be produced from an S.S.B. signal, or any other signal for that matter.

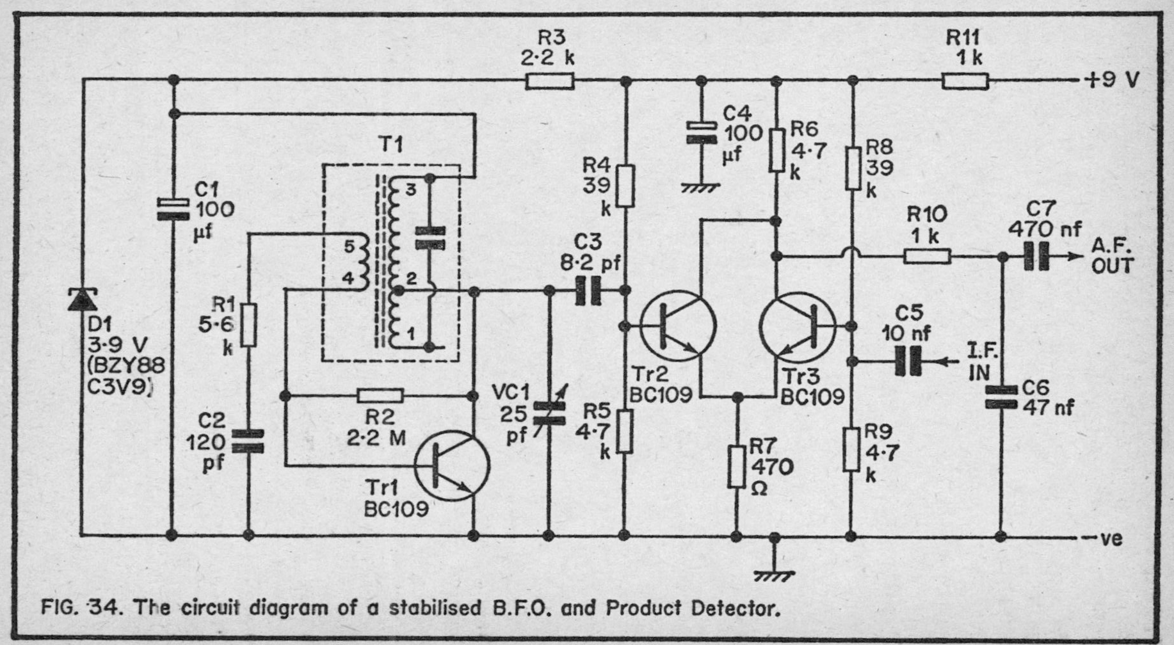

FIG. 34. The circuit diagram of a stabilised B.F.O. and Product Detector.

Tr1 is used as a simple inductive feedback oscillator, and its supply rail potential is stabilised by R3 and D1. VC1 is the B.F.O. frequency control.

Tr2 and Tr3 are the active devices used in the product detector. The B.F.O. signal is coupled to one input of the product detector via C3, and the I.F. output is fed into the other input via C5.

The product detector will not properly demodulate an A.M. transmission, and so it is necessary to incorporate switching from the ordinary diode detector to the product type, in the receiver. One way of doing this is shown in the circuit diagram of Figure 35.

The circuit works normally when S1 is in the A.M. position, and neither the I.F. input or the audio output of the product detector are connected to the rest of the receiver. Also, S1d cuts the supply line to the B.F.O. and product detector circuits.

In the C.W./S.S.B. position S1 connects the I.F. output to the relevant input of the product detector. Note however, that the I.F. output is still connected to the diode detector. S1c switches the volume control into the output circuit of the product detector, and so it no longer forms the load for the diode detector. RX is switched into circuit as the diode load resistor in place of the volume control, and so the biasing of the I.F. amplifier is not significantly affected. Also, the A.G.C. circuitry will still function normally.

If this circuitry is being added to a receiver other than one of those described in this book, there should be no real problems here provided the I.F. output circuitry is of a basically similar type. RX is given a value which is equal to that of the volume control.

One important constructional point must be borne in mind, and that concerns the wiring around S1a. This must either be

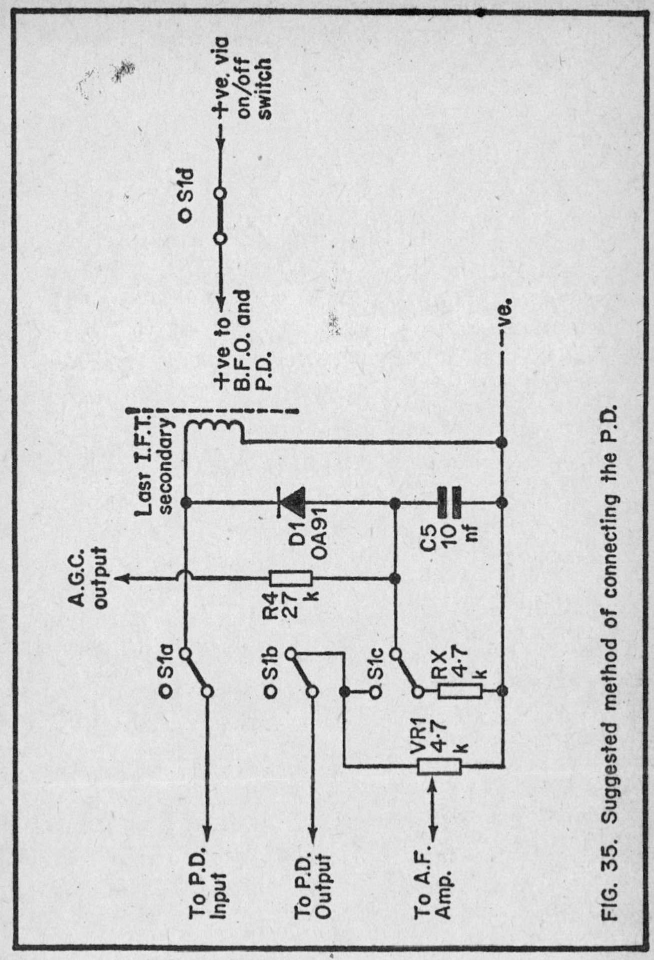

FIG. 35. Suggested method of connecting the P.D.

kept very short or should be screened as otherwise stray feedback over the I.F. amplifier stages is likely to cause problems, and could easily result in this part of the set breaking into violent oscillation.

The only adjustment which needs to be made to the finished unit is that to the core of T1 to align the B.F.O. This is carried out in the usual fashion. T1 is a Denco I.F.T.14/470kHZ component.

Mains P.S.U.

The receivers described in this book can be powered from a fairly large 9 volt battery, such as a PP6, PP7, PP9, or equivalents. This is likely to be rather expensive in the long term though, especially if a number of add-on circuits are incorporated in the receiver, since the current consumption of the set will be quite high. In the long term a mains P.S.U. is likely to be a more economic proposition. The circuit diagram of a stabilised mains power supply unit is shown in Figure 36.

T1 is a mains transformer having a secondary voltage of 12V. and a current rating of 250mA. or more. Its output is fullwave rectified by the bridge rectifier consisting of D1 to D4. C1 smoothes the rough D.C. output of the rectifier network, and FS1 protects the circuit against accidental short circuits at the output.

Tr1 is connected as a conventional emitter follower regulator circuit, and as well as providing a well stabilised output this also contributes considerably to the smoothing of the output. In consequence there is a negligible amount of hum and noise on the output voltage.

Several constructional points should be carefully noted. The magnetic field of the mains transformer can easily induce mains hum into any wire which runs close to it. It can even induce mains hum into the R.F. coils, and so it is a good idea to mount this component as far away from other components as possible. It is also advisable to keep the mains wiring away from any A.F. wiring, as this can also lead to a noticeable level of mains hum on the output signal. Any mains hum that is apparent will almost certainly have been caused by stray pick up as the level of hum on the output of the P.S.U. is not significant.

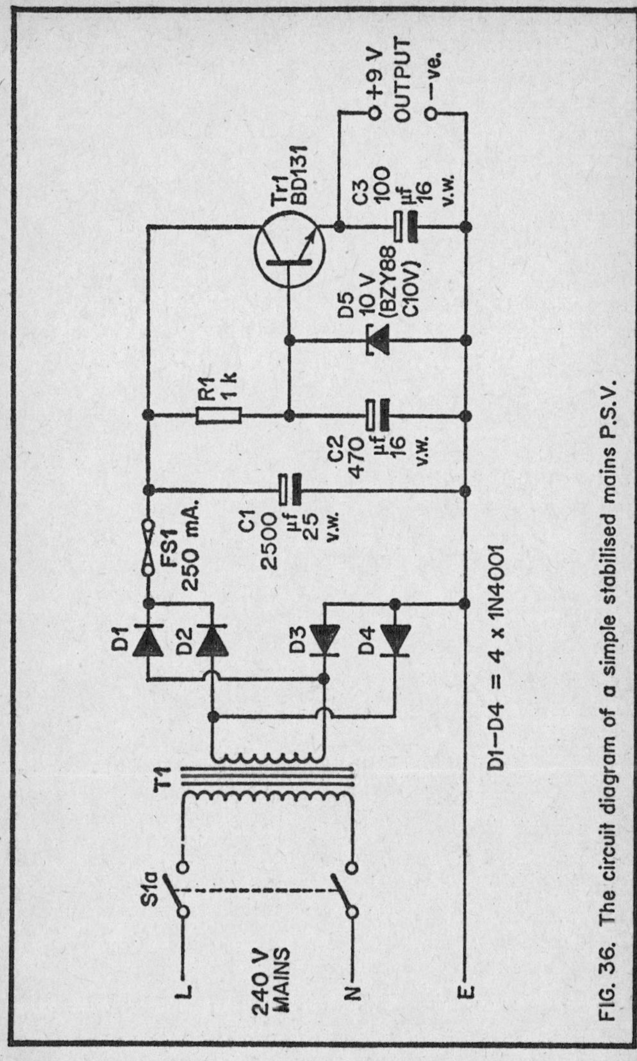

FIG. 36. The circuit diagram of a simple stabilised mains P.S.V.

Tr1 is a power transistor and this is mounted on the metal chassis of the receiver which then acts as a heatsink. This component must be insulated from the chassis using the usual mica washer and plastic bush set. Use a continuity tester to make absolutely sure that this insulation is effective as otherwise the output of the P.S.U. will be short circuited (the metal pad on the underside of the transistor connects to its collector terminal).

For reasons of safety, the receiver must be enclosed in a proper casing which leaves no mains wiring exposed, or within easy access. Also, whether an earth connection is used or not, the chassis of the receiver must be connected to the mains earth.

Crystal Calibrator

Home constructed receivers are easier and more convenient to use if they are equipped with a calibrated tuning dial. If a good quality wide range R.F. signal generator is available, this can be used to supply the calibration signal. An alternative, and probably more popular method of obtaining a calibration signal is to use a device known as a crystal calibration oscillator.

This an oscillator which operates on a single frequency, or perhaps on two or three frequencies, and has a strong harmonic output throughout the S.W. frequency spectrum. This type of circuit usually operates at a fundamental frequency of 1MHZ, with a lower fundamental frequency available if required (usually 100kHZ). Thus such a unit can provide calibration signals in the form of 1MHZ and 100kHZ harmonics. The harmonics are merely signals produced at multiples of the fundamental frequency, and so a 1MHZ calibrator will produce outputs at 2MHZ (second harmonic), 3MHZ (third harmonic), 4MHZ, etc. to beyond 30MHZ. It therefore enables the tuning dial of the receiver to be calibrated at 1MHZ intervals throughout its tuning range. A 100kHZ calibration signal provides additional calibration points at 100kHZ intervals throughout the tuning range of the receiver. Crystal calibrators sometimes have even lower fundamental frequencies, occasionally as low as 10kHZ, but for general coverage receivers such as those featured in this book, 100kHZ is about the practical minimum unless an exceptionally good tuning drive and dial is fitted.

The circuit diagram of a 1MHZ/100kHZ crystal calibration oscillator is shown in Figure 37. Tr1 and Tr2 form the basis of the oscillator circuit, and these are both used as common emitter amplifiers. The output of Tr1 is coupled to the input of Tr2 by way of C1. Tr2 collector and Tr1 base are therefore in phase, and any feedback between the two will be of the positive variety.

The crystal and TC1 are connected between these two points, and a large amount of positive feedback will be produced at the series resonant frequency of the crystal where it exhibits a low impedance. The circuit thus oscillates violently at this frequency producing a high amplitude output which is rich in harmonics.

Rather than use a separate crystal to provide a 100kHZ output signal, the output of the 1MHZ oscillator is instead fed to a CMOS decade counter/divider which divides the 1MHZ signal by ten so that a 100kHZ output is produced.

Using The Calibrator

In order for maximum accuracy to be obtained it is necessary to adjust TC1 to trim the oscillator frequency to precisely 1MHZ. One way of doing this is to connect a lead about 500m.m. long to the 100kHZ output, and then place this lead close to the ferrite aerial of an operating radio which is tuned to B.B.C. Radio 2 on 200kHZ L.W. The second harmonic of the calibrator will produce a heterodyne as it reacts with the 200kHZ carrierwave of Radio 2, but this beat note will be at only a very low frequency (no more than a few tens of HZ). TC1 is adjusted to produce the lowest possible beat note. It should be possible to obtain an output of less than one pulse per second.

It is worth noting that even if the calibrator is not set up in this way, and TC1 is simply adjusted for about half maximum capacitance, the accuracy of the unit should still be greater than is really necessary.

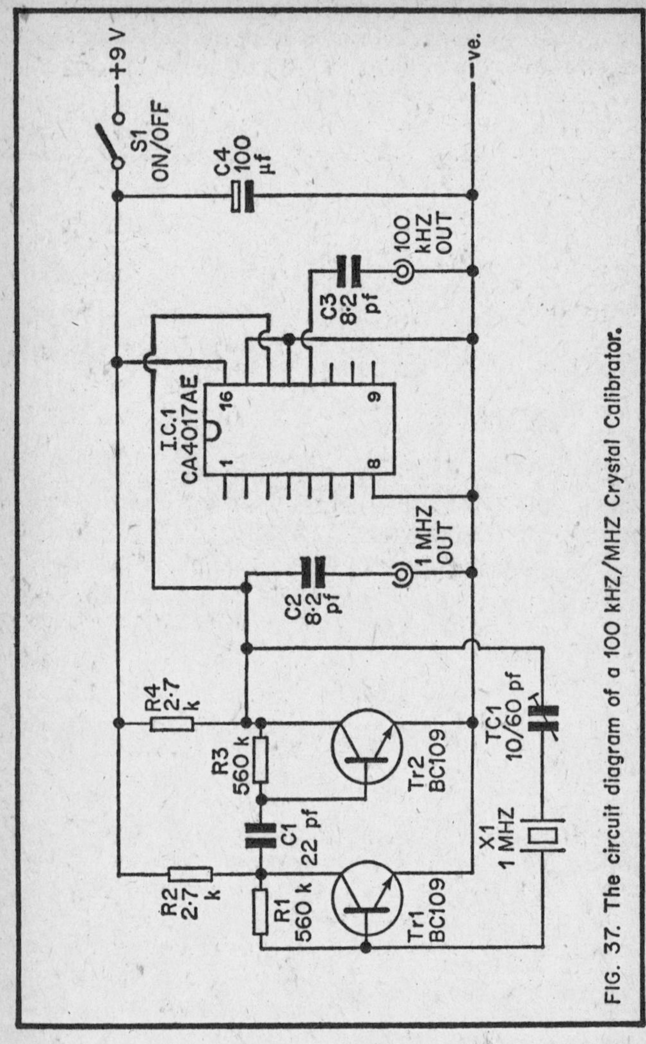

FIG. 37. The circuit diagram of a 100 kHZ/MHZ Crystal Calibrator.

In use it is best to use the 1MHZ output to enable the 1MHZ intervals to be marked in before locating the 100kHZ calibration points. The output of the calibrator is quite strong and there should be no need to couple it direct to the aerial terminal of the receiver. In fact this is not a good idea as it can easily lead to confusing results. It is a good idea to use only a loose coupling between the two units, and it is usually sufficient to connect a lead to the output of the calibrator and then place this near the aerial terminal of the receiver. If a tighter coupling is required (as it might well be at high frequencies), then an insulated lead can be connected to both the output of the calibrator and to the aerial terminal of the receiver. These two leads are then twisted together to form a very low value capacitor..

On Range 5T it might be found rather difficult to identify one harmonic from another as they will be rather closely spaced. One way of achieving this is to tune the set to a signal of known or approximately known frequency. For example, if the set is tuned to an amateur station on the 20 metre amateur band (which extends from 14.0–14.35MHZ) and then coupled to the calibrator, the first marker signal which will be received if the set is tuned lower in frequency must be the 14th harmonic at 14.0MHZ. If the set is then tuned higher in frequency the 15th harmonic at 15MHZ will be received, then the 16th harmonic at 16.0MHZ, and so on. In this way all the harmonics can be identified.

Preselector

A preselector is a tuned R.F. amplifier which is built as an entirely separate unit from the receiver. Apart from increasing the overall gain of a receiving set up it also provides a very useful improvement in image rejection when used in conjunction with a superhet receiver. The circuit diagram of a preselector which uses a cascode arrangement appears in Figure 38.

The aerial signal is coupled to the primary winding of T1, and the secondary winding couples into the g1 terminal of Tr1 which is a dual gate MOSFET. The secondary winding of T1 is

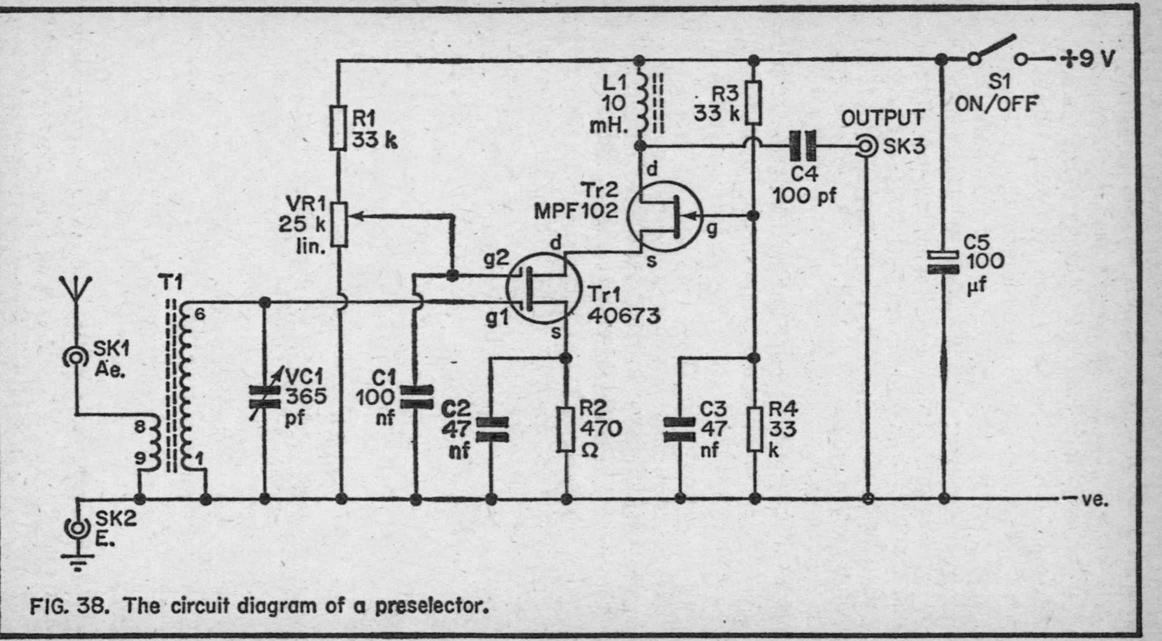

FIG. 38. The circuit diagram of a preselector.

also the tuned winding, and VC1 is the tuning capacitor. Tr1 is used as a common source amplifier, and R2 is its source bias resistor. C2 is the bypass capacitor for the source resistor. VR1 controls the g2 potential of Tr1 and it thus operates as the R.F. gain control.

Tr2 is a Jugfet, and it is used in the common gate mode. R3 and R4 bias the gate to about half the supply rail potential and C3 is the earth return capacitor for Tr2 gate. Since the gate of Tr2 is held at a fixed potential, as the drain of Tr1 swings positive and negative it will take Tr2 source with it, and this results in a variation of the g – s voltage of Tr2. This results in Tr2 conducting more and less heavily as its source terminal goes positive and negative. This causes an amplified signal to be generated in the drain circuit of Tr2, and this signal is developed across L1.

C4 provides D.C. blocking at the output and C5 is a supply decoupling capacitor. S1 is an ordinary on/off switch.

T1 is a Denco Dual Purpose (D.P.) coil, and as was the case with the S.W. receivers described earlier, three coils are needed to cover the entire S.W. frequency spectrum. These are ranges 3, 4, and 5, and the coils are colour coded BLUE. L1 is a Repanco ferrite cored R.F. choke type CH4, although any similar R.F. choke is suitable.

In use the output of the preselector is coupled to the aerial socket of the receiver via a screened (coaxial) cable which should be kept as short as possible so that losses here are minimised. The core of T1 is given a setting that enables VC1 to peak received signals at any setting of the tuning control on the receiver. This adjustment must be made to each of the three coils.

It has been assumed above that the receiver is a home made type using Denco coils, but, of course, the preselector can be used with just about any S.W. receiver. The cores of the three coils are then given any settings which do not result in there being any gaps in the frequency coverage. There is considerable overlap between the frequency coverage of adjacent ranges, and so the core settings are not particularly critical.

It should perhaps be pointed out that there is one type of set which the unit cannot be used with, and that is sets of the old A.C./D.C. type which are mains powered and do not incorporate a mains isolation transformer. With this type of set the chassis is connected to one side of the mains, and since the chassis of the preselector and receiver are connected via the coaxial cables outer braiding, the chassis of the preselector would be at mains potential if it were used with a set of this type. There would then be a danger of the user receiving a severe electric shock.

The gain of the circuit is quite high, particularly on the L.F. bands, and so the R.F. gainccontrol should not simply be set at maximum and left there. This could easily result in the set being continually overloaded. What was said in Chapter 2 about using internal R.F. amplifiers also pertains to the use of preselectors.

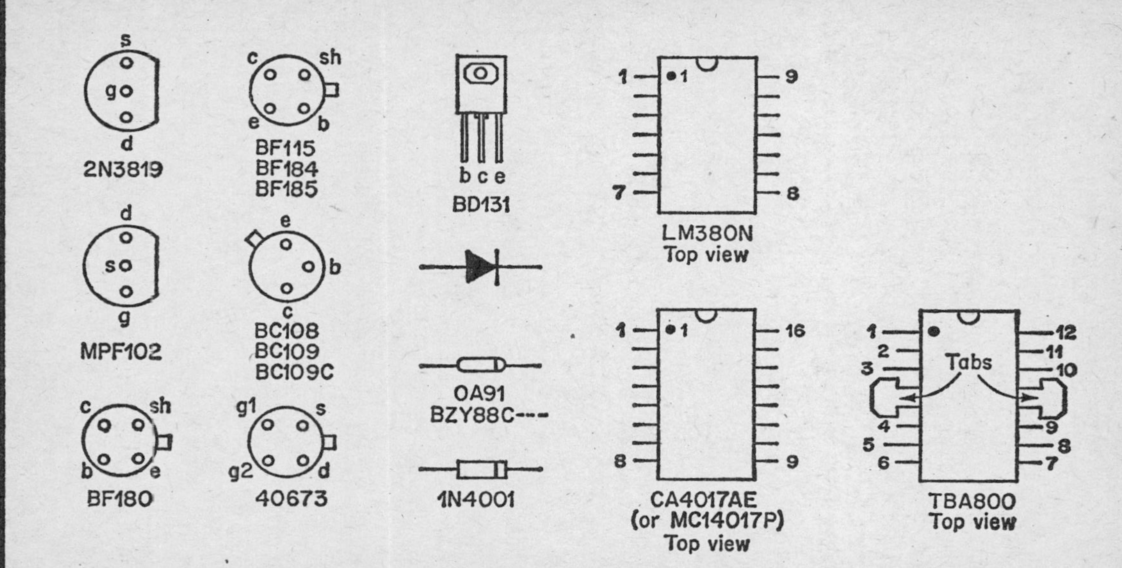

FIG. 39. Semiconductor leadout data. Transistors are viewed onto base.

SPECIAL NOTES FOR OVERSEAS READERS

1. If you should experience any difficulty in obtaining "DENCO" coils locally, then these can always be ordered direct from the manufacturers whose address is shown below:–

DENCO (Clacton) LTD
355/9 Old Road
Clacton-on-Sea
Essex. CO15 3RH
England
Tel.: Clacton 22807

Notes

Notes

Notes

Notes

Notes

Notes

OTHER BOOKS BY R.A. PENFOLD PUBLISHED BY BERNARDS (PUBLISHERS) LTD

No.222 SOLID STATE SHORT WAVE RECEIVERS FOR BEGINNERS
I.S.B.N. 0 900162 62 7 **Price 95p**

* There is a strange fascination in listening to a broadcast which has been transmitted from a station that may be many thousands of miles away across the other side of the world. This has helped to make shortwave listening one of the most popular and interesting branches of electronics and for this reason, is an excellent way of capturing the interest of a beginner. In fact, very many enthusiasts and "Hams" have been introduced to the hobby in this way.

* In this book the author has designed and developed several modern solid state shortwave receiver circuits that will give a fairly high level of performance, despite the fact that they use relatively few and inexpensive components.

* An ideal book for the beginnier or inexperienced enthusiast.

No.223 50 PROJECTS USING IC CA3130
I.S.B.N. 0 900162 65 1 **Price 95p**

* The CA3130 is currently one of the more advanced operational amplifiers that is available to the home constructor. This means that it is often capable of a higher level of performance than many other devices and that it often needs fewer ancillary components.

* In this book the author has designed and developed a number of interesting and useful projects which are divided into five general categories.

I — Audio Projects
II — R.F. Projects
III — Test Equipment
IV — Household Projects
V — Miscellaneous Projects

* An ideal book for both beginner and more advanced enthusiastis alike.

No.224 50 CMOS IC PROJECTS
I.S.B.N. 0 900162 64 3 **Price 95p**

* CMOS IC's are probably the most versatile range of digital devices for use by the amateur enthusiast. They are suitable for an extraordinarily wide range of applications and are now also some of the most inexpensive and easily available types of I.C.

* In this book the author has designed and developed a number of interesting and useful projects which are divided into four general categories.

I — Multivibrators
II — Amplifiers and Oscillators
III — Trigger Devices
IV — Special Devices

* A must for the library of any Electronic Enthusiast.

Please note overleaf is a list of other titles that are available in our range of Radio and Electronic Books.

These should be available from all good Booksellers, Radio Component Dealers and Mail Order Companies.

However, should you experience difficulty in obtaining any title in your area, then please write directly to the publisher enclosing payment to cover the cost of the book plus adequate postage.

BABANI PRESS & BERNARDS (PUBLISHERS) LTD
THE GRAMPIANS
SHEPHERDS BUSH ROAD
LONDON W6 7NF
ENGLAND

BERNARDS & BABANI PRESS RADIO AND ELECTRONICS BOOKS

BP1	First Book of Transistor Equivalents and Substitutes	40p
BP2	Handbook of Radio, TV and Ind. & Transmitting Tube & Valve Equiv.	60p
BP3	Handbook of Tested Transistor Circuits	40p
BP4	World's Short, Med. & LW FM & TV Broadcasting Stations Listing	60p
BP5	Handbook of Simple Transistor Circuits	35p
BP6	Engineers and Machinists Reference Tables	30p
BP7	Radio and Electronic Colour Codes and Data Chart	15p
BP8	Sound and Loudspeaker Manual	50p
BF9	38 Practical Tested Diode Circuits for the Home Constructor	35p
BP10	Modern Crystal and Transistor Set Circuits for Beginners	35p
BP11	Practical Transistor Novelty Circuits	40p
BP12	Hi-Fi, P.A., Guitar & Discotheque Amplifier Handbook	75p
BP13	Electronic Novelties for the Motorist	50p
BP14	Second Book of Transistor Equivalents	95p
BP15	Constructors Manual of Electronic Circuits for the Home	50p
BP16	Handbook of Electronic Circuits for the Amateur Photographer	60p
BP17	Radio Receiver Construction Handbook using IC's and Transistors	60p
BP18	Boys and Beginners Book of Practical Radio and Electronics	60p
BP22	79 Electronic Novelty Circuits	75p
BP23	First Book of Practical Electronic Projects	75p
BP24	52 Projects using IC741 (or Equivalents)	75p
BP25	How to Build Your Own Electronic and Quartz Controlled Watches & Clocks	85p
BP26	Radio Antenna Handbook for Long Distance Reception & Transmission	85p
BP27	Giant Chart of Radio Electronic Semiconductor & Logic Symbols	60p
BP28	Resistor Selection Handbook (International Edition)	60p
BP29	Major Solid State Audio Hi-Fi Construction Projects	85p
BP30	Two Transistor Electronic Projects	85p
BP31	Practical Electrical Re-wiring & Repairs	85p
BP32	How to Build Your Own Metal and Treasure Locators	85p
BF83	Electronic Calculator Users Handbook	95p
BP34	Practical Repair & Renovation of Colour TV's	95p
BP35	Handbook of IC Audio Preamplifier & Power Amplifier Construction	95p
BP36	50 Circuits using Germanium, Silicon and Zener Diodes	75p
BP37	50 Projects using Relays, SCRs and TRIACs	£1-10p
100	A Comprehensive Radio Valve Guide – Book 1	40p
121	A Comprehensive Radio Valve Guide – Book 2	40p
126	Boys Book of Crystal Sets	25p
129	Universal Gram-Motor Speed Indicator (Combined 50 & 60 c/s model)	10p
138	How to Make Aerials for TV (Band 1-2-3)	25p
143	A Comprehensive Radio Valve Guide – Book 3	40p
150	Practical Radio Inside Out	40p
157	A Comprehensive Radio Valve Guide – Book 4	40p
160	Coil Design and Construction Manual	50p
161	Radio TV and Electronics Data Book	60p
170	Transistor Circuits for Radio Controlled Models	40p
177	Modern Transistor Circuits for Beginners	40p
178	A Comprehensive Radio Valve Guide – Book 5	40p
183	How to Receive Foreign TV Programmes on your Set by Simple Mods.	35p
196	AF–RF Reactance – Frequency Chart for Constructors	15p
200	Handbook of Practical Electronic Musical Novelties	50p
201	Practical Transistorised Novelties for Hi-Fi Enthusiasts	35p
202	Handbook of Integrated Circuits (IC's) Equivalents and Substitutes	75p
203	IC's and Transistor Gadgets Construction Handbook	60p
204	Second Book of Hi-Fi Loudspeaker Enclosures	60p
205	First Book of Hi-Fi Loudspeaker Enclosures	60p
206	Practical Transistor Circuits for Modern Test Equipment	60p
207	Practical Electronic Science Projects	75p
208	Practical Stereo and Quadrophony Handbook	75p
209	Modern Tape Recording Handbook	75p
210	The Complete Car Radio Manual	75p
211	First Book of Diode Characteristics Equivalents and Substitutes	95p
213	Electronic Circuits for Model Railways	85p
214	Audio Enthusiasts Handbook	85p
215	Shortwave Circuits and Gear for Experimenters and Radio Hams	85p
216	Electronic Gadgets and Games	85p
217	Solid State Power Supply Handbook	85p
218	Build Your Own Electronic Experimenters Laboratory	85p
219	Solid State Novelty Projects	85p
220	Build Your Own Solid State Hi-Fi and Audio Accessories	85p
221	28 Tested Transistor Projects	95p
222	Solid State Short Wave Receivers for Beginners	95p
223	50 Projects using IC CA3130	95p
224	50 CMOS IC Projects	95p
RCC	Resistor Colour Code Disc Calculator	10p

The Grampians, Shepherds Bush Road, London W6 7NF, England

Telephone: 01-603 2581/7296